# HYPOELLIPTIC BOUNDARY-VALUE PROBLEMS

# PURE AND APPLIED MATHEMATICS

*A Program of Monographs, Textbooks and Lecture Notes*



Contributions to *Lecture Notes in Pure and Applied Mathematics* are reproduced by direct photography of the author's typewritten manuscript. Potential authors are advised to submit preliminary manuscripts for review purposes. After acceptance, the author is responsible for preparing the final manuscript in camera-ready form, suitable for direct reproduction. Marcel Dekker, Inc. will furnish instructions to authors and special typing paper. Sample pages are reviewed and returned with our suggestions to assure quality control and the most attractive rendering of your manuscript. The publisher will also be happy to supervise and assist in all stages of the preparation of your camera-ready manuscript.

# LECTURE NOTES
# IN PURE AND APPLIED MATHEMATICS

1. *N. Jacobson,* Exceptional Lie Algebras
2. *L.-Å. Lindahl and F. Poulsen,* Thin Sets in Harmonic Analysis
3. *I. Satake,* Classification Theory of Semi-Simple Algebraic Groups
4. *F. Hirzebruch, W. D. Newmann, and S. S. Koh,* Differentiable Manifolds and Quadratic Forms
5. *I. Chavel,* Riemannian Symmetric Spaces of Rank One
6. *R. B. Burckel,* Characterization of C(X) Among Its Subalgebras
7. *B. R. McDonald, A. R. Magid, and K. C. Smith,* Ring Theory: Proceedings of the Oklahoma Conference
8. *Y.-T. Siu,* Techniques of Extension of Analytic Objects
9. *S. R. Caradus, W. E. Pfaffenberger, and B. Yood,* Calkin Algebras and Algebras of Operators on Banach Spaces
10. *E. O. Roxin, P.-T. Liu, and R. L. Sternberg,* Differential Games and Control Theory
11. *M. Orzech and C. Small,* The Brauer Group of Commutative Rings
12. *S. Thomeier,* Topology and Its Applications
13. *J. M. López and K. A. Ross,* Sidon Sets
14. *W. W. Comfort and S. Negrepontis,* Continuous Pseudometrics
15. *K. McKennon and J. M. Robertson,* Locally Convex Spaces
16. *M. Carmeli and S. Malin,* Representations of the Rotation and Lorentz Groups: An Introduction
17. *G. B. Seligman,* Rational Methods in Lie Algebras
18. *D. G. de Figueiredo,* Functional Analysis: Proceedings of the Brazilian Mathematical Society Symposium
19. *L. Cesari, R. Kannan, and J. D. Schuur,* Nonlinear Functional Analysis and Differential Equations: Proceedings of the Michigan State University Conference
20. *J. J. Schäffer,* Geometry of Spheres in Normed Spaces
21. *K. Yano and M. Kon,* Anti-Invariant Submanifolds
22. *W. V. Vasconcelos,* The Rings of Dimension Two
23. *R. E. Chandler,* Hausdorff Compactifications
24. *S. P. Franklin and B. V. S. Thomas,* Topology: Proceedings of the Memphis State University Conference
25. *S. K. Jain,* Ring Theory: Proceedings of the Ohio University Conference
26. *B. R. McDonald and R. A. Morris,* Ring Theory II: Proceedings of the Second Oklahoma Conference
27. *R. B. Mura and A. Rhemtulla,* Orderable Groups
28. *J. R. Graef,* Stability of Dynamical Systems: Theory and Applications
29. *H.-C. Wang,* Homogeneous Banach Algebras
30. *E. O. Roxin, P.-T. Liu, and R. L. Sternberg,* Differential Games and Control Theory II
31. *R. D. Porter,* Introduction to Fibre Bundles
32. *M. Altman,* Contractors and Contractor Directions Theory and Applications
33. *J. S. Golan,* Decomposition and Dimension in Module Categories
34. *G. Fairweather,* Finite Element Galerkin Methods for Differential Equations
35. *J. D. Sally,* Numbers of Generators of Ideals in Local Rings
36. *S. S. Miller,* Complex Analysis: Proceedings of the S.U.N.Y. Brockport Conference
37. *R. Gordon,* Representation Theory of Algebras: Proceedings of the Philadelphia Conference
38. *M. Goto and F. D. Grosshans,* Semisimple Lie Algebras
39. *A. I. Arruda, N. C. A. da Costa, and R. Chuaqui,* Mathematical Logic: Proceedings of the First Brazilian Conference
40. *F. Van Oystaeyen,* Ring Theory: Proceedings of the 1977 Antwerp Conference

41. *F. Van Oystaeyen and A. Verschoren,* Reflectors and Localization: Application to Sheaf Theory
42. *M. Satyanarayana,* Positively Ordered Semigroups
43. *D. L. Russell,* Mathematics of Finite-Dimensional Control Systems
44. *P.-T. Liu and E. Roxin,* Differential Games and Control Theory III: Proceedings of the Third Kingston Conference, Part A
45. *A. Geramita and J. Seberry,* Orthogonal Designs: Quadratic Forms and Hadamard Matrices
46. *J. Cigler, V. Losert, and P. Michor,* Banach Modules and Functors on Categories of Banach Spaces
47. *P.-T. Liu and J. G. Sutinen,* Control Theory in Mathematical Economics: Proceedings of the Third Kingston Conference, Part B
48. *C. Byrnes,* Partial Differential Equations and Geometry
49. *G. Klambauer,* Problems and Propositions in Analysis
50. *J. Knopfmacher,* Analytic Arithmetic of Algebraic Function Fields
51. *F. Van Oystaeyen,* Ring Theory: Proceedings of the 1978 Antwerp Conference
52. *B. Kedem,* Binary Time Series
53. *J. Barros-Neto and R. A. Artino,* Hypoelliptic Boundary-Value Problems

*Other Volumes in Preparation*

# HYPOELLIPTIC BOUNDARY-VALUE PROBLEMS

**J. Barros-Neto**
Department of Mathematics
Rutgers University
New Brunswick, New Jersey

**Ralph A. Artino**
Department of Mathematics
The City College
New York, New York

MARCEL DEKKER, INC.    New York and Basel

Library of Congress Cataloging in Publication Data

Barros-Neto, J.
Hypoelliptic boundary-value problems.

(Lecture notes in pure and applied mathematics ; 53)
Bibliography: p.
Includes index.
1. Boundary-value problems--Numerical solutions.
2. Differential equations, Hypoelliptic--Numerical solutions. I. Artino, Ralph A. joint author. II. Title.
QA379.B37 515.3'5 79-29732
ISBN 0-8247-6886-8



MARCEL DEKKER, INC.

270 Madison Avenue, New York, New York 10016

Current printing (last digit):

10 9 8 7 6 5 4 3 2 1

PRINTED IN THE UNITED STATES OF AMERICA

# PREFACE

This book is a survey of the theory of hypoelliptic boundary-value problems in the constant coefficients case. The theory was initiated by Hormander in his paper [11], where he gave a necessary and sufficient condition for solutions of a homogeneous boundary-value problem to be $C^{\infty}$ up to the boundary of the domain. His condition, of an algebraic nature, was formulated in terms of the behavior near infinity of the zeros of the so called *characteristic function* of the boundary-value problem. Roughly, Hormander's condition is similar to the algebraic condition that characterizes hypoelliptic partial differential operators with constant coefficients.

Later on, Barros-Neto [6] gave another characterization of the same problem based on regularity properties of the fundamental kernels associated to the boundary-value problem under consideration. With the help of such kernels, a parametrix of the boundary-value problem can be constructed.

In his dissertation, Artino [1] investigated Gevrey regularity, up to the boundary, for solutions of hypoelliptic boundary-value problems. By refining Hormander's estimates, he was able to prove Gevrey regularity, up to the boundary, and to show that the corresponding kernels belonged to suitable Gevrey classes. Artino also studied, in his paper [2], the case of semielliptic boundary-value problems where the differential operators involved had mixed degrees of homogeneity. Artino's method proved to be powerful enough to distinguish, in the case of general hypoelliptic boundary-value problems, the different degrees of Gevrey regularity, with respect to the different variables. This was done in [7].

Elliptic operators are characterized in terms of properties of the principal part of the symbol of the operator. In the same way, elliptic boundary-value problems can be characterized in terms of the *principal part* of the *characteristic function* of the boundary-value problem. This fact was observed by Hormander in his paper [11]. It turns out that, when dealing with semielliptic boundary-value problems, it is possible to define what one may call the principal part of the characteristic function and characterize the boundary-value problem in terms of properties of such a principal part. This was carried out in a joint paper by Artino and Barros-Neto [3].

The plan of this book is as follows. In Chapter 1, we review, briefly, the definition and main properties of hypoelliptic operators and prove necessary and sufficient conditions for hypoellipticity and d-hypoellipticity. Most of the results of Chapter 1 can be found in [12] and [16].

Chapter 2 is devoted to hypoelliptic boundary-value problems. The characteristic function of an hypoelliptic boundary-value problem is defined and its main properties are proved. Necessary and sufficient conditions for $C^\infty$ (resp. Gevrey) regularity of solutions are obtained. We construct the fundamental kernels of the boundary-value problem and prove regularity outside of the origin. A parametrix of the hypoelliptic problem is then obtained.

In Chapter 3, we discuss semielliptic boundary-value problems. First, we consider the case where all the operators are semihomogeneous and prove that the kernels split into a sum of a semihomogeneous distribution and a semihomogeneous polynomial multiplied by a suitable factor. Next, we consider semielliptic problems, in general, and prove an algebraic characterization in terms of the principal part of the characteristic function. Our condition generalizes Hormander's condition in the case of elliptic boundary-value problems.

This book is intended for graduate students interested in the theory of partial differential operators. It deals mainly with regularity results for solutions of hypoelliptic boundary-value problems. Since elliptic boundary problems are a particular case of hypoelliptic problems, our results contain, as a particular case, the classical results about elliptic operators with constant coefficients. A student with some knowledge of distribution theory and the classical theory of partial differential operators should encounter no difficulties in reading this book. We also hope that the material presented here will serve as a motivation to further studies in the field of partial differential equations.

Ralph A. Artino
J. Barros-Neto

CONTENTS

# HYPOELLIPTIC BOUNDARY-VALUE PROBLEMS

# CHAPTER 1
# HYPOELLIPTIC DIFFERENTIAL OPERATORS

## 1.1 DEFINITION, MAIN PROPERTIES

We shall deal with partial differential operators

$$P = P(D) = \sum_{|p| \leq m} a_p D^p \tag{1.1}$$

with constant coefficients $a_p \in \mathbb{C}$ and degree $\geq 1$, such that all distributions solutions of the equation

$$Pu = f$$

are always smooth functions whenever f is a smooth function.

Definition 1.1. *We say that the differential operator* $P(D)$ *is hypoelliptic if, for every open set* $\Omega \subset \mathbb{R}^n$ *and every distribution* $T \in \mathcal{D}'(\Omega)$, $PT \in C^\infty(\Omega)$ implies $T \in C^\infty(\Omega)$.

Let $\zeta = (\zeta_1,\ldots,\zeta_n) \in \mathbb{C}^n$ with $\zeta_j = \xi_j + i\eta_j$, $1 \leq j \leq n$. The polynomial

$$P(\zeta) = \sum_{|p| \leq m} a_p \zeta^p \tag{1.2}$$

is called the *characteristic polynomial* (or *symbol*) of $P(D)$. Denote by

$$N = \{\zeta \in \mathbb{C}^n : P(\zeta) = 0\} \tag{1.3}$$

the variety of zeros of $P(\zeta)$. For every $\xi \in \mathbb{R}^n$, let

$$d(\xi,N) = \inf_{\xi\varepsilon N} |\xi-\zeta|$$

be the distance from $\xi$ to N.

Theorem 1.1. *Let $P(\zeta)$ be a constant coefficient polynomial. The following properties are equivalent:*

$(H)_1$ $\zeta \in N$, $|\zeta| \to +\infty$ *implies* $|\mathrm{Im}\zeta| \to +\infty$;

$(H)_2$ $\xi \in \mathbb{R}^n$, $|\xi| \to \infty$ *implies* $d(\xi,N) \to +\infty$;

$(H)_3$ *for all n-tuples* $p = (p_1,\ldots,p_n)$ *with* $|p| \neq 0$, $\xi \in \mathbb{R}^n$,

$$|\xi| \to +\infty \quad \textit{implies} \quad \frac{|P^{(p)}(\xi)|}{|1+ P(\xi)|} \to 0.$$

The equivalence between $(H)_1$ and $(H)_2$ is easily established, while the equivalence between $(H)_2$ and $(H)_3$ is a consequence of the following lemma whose proof is found in [16].

Lemma 1.1. *Let $m \geq 1$ be an integer. There is a constant $C_m > 0$ such that for all polynomials P of degree $\leq m$, we have*

$$C_m^{-1} \leq d(\xi,N) \sum_{|p|\neq 0} \frac{|P^{(p)}(\xi)|}{|P(\xi)|}^{1/|p|} \leq C_m,$$

*for all $\xi \in \mathbb{R}^n$ such that $P(\xi) \neq 0$.*

It can be shown that condition $(H)_1$ is a necessary one in order that a differential operator be hypoelliptic. More precisely, we have the following result

Theorem 1.2. *Let $P = P(D)$ be a constant coefficient partial differential operator. Suppose that for some open subset $\Omega \subset \mathbb{R}^n$, every $u \in \mathcal{D}'(\Omega)$ such that $Pu = 0$ belongs to $C^\infty(\Omega)$. Then, property $(H)_1$ holds.*

*Proof.* By our assumption

$$H_P(\Omega) = \{u \in L^\infty_{loc}(\Omega) : P(D)u = 0\}$$

is a linear subspace of $C^\infty(\Omega)$. It is easy to see that $H_P(\Omega)$ is closed in

$L^\infty_{loc}(\Omega)$ hence, equipped with the topology induced by $L^\infty_{loc}(\Omega)$, $H_P(\Omega)$ is a Frechet space (i.e. metrizable and complete). Let $\Omega'$ be relatively compact open set such that $\bar{\Omega}' \subset \Omega$ and denote by $T|_{\Omega'}$ the restriction to $\Omega'$ of every distribution $T \in \mathcal{D}'(\Omega)$. If $\Delta$ denotes the Laplace operator in $\mathbb{R}^n$, define the linear map

$$L: u \in H_P(\Omega) \to \Delta u|_{\Omega'} \in L^\infty(\Omega').$$

This linear map is closed hence continuous from $H_P(\Omega)$ into $L^\infty(\Omega')$. Therefore, there is a compact subset $K \subset \Omega$ and a constant $C > 0$ such that

$$||Lu||_{L^\infty(\Omega')} \le C\, ||X_K u||_{L^\infty(\Omega)}, \qquad \forall\, u \in H_P(\Omega),$$

where $X_K$ is the characteristic function of $K$. Let $u(x) = e^{i<x,\zeta>}$ with $P(\zeta) = 0$ and set

$$\delta = \sup_{x\in K} |x| \quad \text{and} \quad \delta' = \sup_{x\in\Omega'} |x|$$

Replacing the above we derive the inequality

$$|\zeta^2| \le C\, e^{(\delta+\delta')|\mathrm{Im}\zeta|}$$

which implies $(H)_1$. Q.E.D.

We observe that if $P(D)$ is hypoelliptic, the hypothesis of Theorem 1.2 holds, hence $P(D)$ satisfies condition $(H)_1$ and, by Theorem 1.1, it satisfies each one of the equivalent conditions $(H)_1$, $(H)_2$ and $(H)_3$. As we shall see later, any one of these equivalent conditions is a sufficient one for hypoellipticity. We can then state the following result.

Theorem 1.3. *A partial differential operator* $P(D)$ *is hypoelliptic if and only if in some open subset all distributions solutions of the homogeneous equation* $P(D)u = 0$ *are* $C^\infty$ *functions.*

Because of Theorem 1.2, we often say that a polynomial $P(\zeta)$ with constant coefficients is hypoelliptic if it satisfies any one of the equivalent conditions $(H)_1$, $(H)_2$ and $(H)_3$. From now on we shall refer to condition $(H)$ as any one of the conditions $(H)_1$, $(H)_2$ and $(H)_3$.

Condition $(H)_1$ immediately implies that the set of real zeros of $P$, namely

$$\{\xi \in \mathbb{R}^n : P(\xi) = 0\} = N \cap \mathbb{R}^n$$

is a compact subset of $\mathbb{R}^n$.

Theorem 1.4. *Let* $P(\zeta)$ *be a constant coefficient polynomial, let* $N$ *be its variety of zeros and let* $d$ *be a real number* $\geq 1$. *The following are equivalent conditions:*

$(dH)_1$ *there is a constant* $C > 0$ *such that* $|\zeta|^{1/d} \leq C(1+|\mathrm{Im}\zeta|)$, *for all* $\zeta \in N$;

$(dH)_2$ *there is a constant* $C' > 0$ *such that* $|\xi|^{1/d} \leq C'(1+d(\xi,N))$, *for all* $\xi \in \mathbb{R}$;

$(dH)_3$ *there is a constant* $C'' > 0$ *such that* $|\xi|^{|p|/d}|P^{(p)}(\xi)| \leq C''(1+|P(\xi)|)$, *for all* $p \in N^n$ *and all* $\xi \in \mathbb{R}^n$.

As in Theorem 1.1, the equivalence between $(dH)_1$ and $(dH)_2$ is easily proved, while to prove $(dH)_2 \Leftrightarrow (dH)_3$ one uses Lemma 1.1 (see Section 1.4). The following remarks are in order.

Remark 1. Condition $(dH)_1$ is obviously equivalent to each of the following conditions:

$(dH)_{1'}$ there is a constant C such that $|\mathrm{Re}\zeta|^{1/d} \leq C(1+|\mathrm{Im}\zeta|)$, for all $\zeta \in N$;

$(dH)_{1''}$ there is a constant C such that $|\mathrm{Re}\zeta| \leq C(1+|\mathrm{Im}\zeta|^d)$, for all $\zeta \in N$.

(we are using the same C to denote different constants).

Remark 2. Condition $(dH)_2$, say, implies that d must be $\geq 1$. Indeed, if $(dH)_2$ holds we have

$$|\xi|^{1-d} \leq C^d \left(\frac{1+d(\xi,N)}{|\xi|}\right)^d, \text{ for all } 0 \neq \xi \in \mathbb{R}^n$$

Let $\zeta \in N$, then $d(\xi,N) \leq |\xi-\zeta|$. We get

$$|\xi|^{1-d} \leq C^d \left(\frac{1+|\xi-\zeta|}{|\xi|}\right)^d$$

As $|\xi| \to +\infty$, the right hand side remains bounded. Thus $(dH)_2$ holds only if $d \geq 1$.

Remark 3. Condition $(dH)_2$ implies trivially condition $(H)_2$. Conversely, $(H)_2$ implies $(dH)_2$. This is a deeper result whose proof uses the Seidenberg-Tarski excision theorem. More precisely, we have the following result [12,16]. As before, we shall refer to condition (dH) as any one of the equivalent conditions $(dH)_1$, $(dH)_2$ and $(dH)_3$.

Theorem 1.5. *Let* $P(\zeta)$ *be a hypoelliptic polynomial. There is a real number* d *such that condition* (dH) *holds. Moreover, the numbers* d *for which condition* (dH) *is valid form a closed half line* $[d_0+\infty)$ *with* $d_0$ *a rational number* $\geq 1$.

*Proof.* Let A be the set of points $(\xi,\alpha,\beta,\tau,\delta) \in \mathbb{R}^n \times \mathbb{R}^n \times \mathbb{R}^n \times \mathbb{R} \times \mathbb{R}$ defined by the following equations and inequalities

$$P(\alpha+i\beta) = 0, \quad \tau > 0, \quad |\xi-\alpha|^2 + |\beta|^2 \leq 1/\tau^2, \quad \delta > 0, \quad |\xi|\delta = 1$$

The set A is semialgebraic [16,pg. 499]. By the Seidenberg-Tarski theorem [16], the image B of A by the projection $(\xi,\alpha,\beta,\tau,\delta) \to (\xi,\tau,\delta)$ is also semialgebraic. It is easily seen that B is defined by the inequalities

$$\tau > 0, \quad d(\xi,N) \leq 1/\tau, \quad \delta > 0, \quad |\xi|\delta = 1$$

If condition $(H)_2$ holds, there is a $\delta_0 > 0$ such that if $0 < \delta < \delta_0$ and $|\xi| = 1/\delta$ we have $d(\xi,N) > 0$. For all such $\delta$ the function

$$\tau(\delta) = \sup_{|\xi|\delta=1} \frac{1}{d(\xi,N)}$$

is well defined and continuous. The image C of B by the projection $(\xi,\tau,\delta) \to (\tau,\delta)$ is a semialgebraic set. Moreover, one can show that there is $\delta_0 > 0$ such that $(\tau(\delta),\delta)$ varies on the boundary of C when $\delta$ belongs to the interval $(0,\delta_0)$. Shrinking this interval, if necessary, it follows that $(\tau(\delta),\delta)$ satisfy an equation $Q(\tau(\delta),\delta) = 0$, with Q a suitable polynomial in two variables. Then, $\tau(\delta)$ admits a Puiseux expansion in some neighborhood of 0 in the complex $\delta$-plane

$$\tau(\delta) = a_k\,(\delta^{1/q})^k + a_{k+1}\,(\delta^{1/q})^{k+1} + \ldots$$

where q is an integer > 0 and k an integer. We can assume that $a_k \neq 0$ and, by choosing for $\delta^{1/q}$ the branch which is > 0 for $\delta > 0$, we can assume

$a_k > 0$. By condition $(H)_2$, $\tau(\delta) \to 0$ as $\delta \to 0$, hence we must have $k > 0$. Thus, near $\delta = 0$, we have

$$\tau(\delta) = a_k\ \delta^{k/q}(1+0(\delta^{1/q}))$$

which implies

$$\tau(\delta) \leq c_1\ \delta^{k/q}, \text{ for } \delta \text{ small}$$

Hence, it follows that

$$\inf_{|\xi|\delta=1} d(\xi,N) = 1/\tau(\sigma) \geq c_2\ |\xi|^{k/q}, \text{ for } |\xi| \text{ large}$$

This last inequality implies $(dH)_2$ with $d = q/k$ and also the fact that the set of numbers d for which $(dH)_2$ holds is a closed half line with a rational number as lower limit. Q.E.D.

In summary, conditions (H) and (dH) are all necessary ones, in order that a partial differential operator with constant coefficients be hypoelliptic. As we shall see, they are also sufficient conditions for hypoellipticity. Before discussing this point let us introduce the notion of d-hypoellipticity.

## 1.2 GEVREY CLASSES AND d-HYPOELLIPTICITY

Definition 1.2. *Let* $\Omega$ *be an open subset of* $\mathbb{R}^n$ *and let* d *be a non-negative real number. We say that a function* $u \in C^\infty(\Omega)$ *belongs to the Gevrey class* d *in* $\Omega$ *if, to every compact set* $K \subset \Omega$, *there is a constant* $C = C(K,u)$ *such that*

$$\sup_{x\in K} |D^p u(x)| \leq C^{|p|+1}(|p|!)^d, \qquad \forall\ p \in N^n \tag{1.4}$$

*We denote by* $\Gamma^d(\Omega)$ *the linear subspace of* $C^\infty(\Omega)$ *of all functions of Gevrey class* d.

When $d = 1$, $\Gamma^1(\Omega)$ coincides with the set of all analytic functions in $\Omega$, usually denoted by $A(\Omega)$. It is well known that, when $d < 1$, the elements of $\Gamma^d(\Omega)$ are analytic functions. When $d > 1$, $\Gamma^d(\Omega)$ contains functions which are not analytic.

Definition 1.3. *We say that a partial differential operator* $P(D)$ *is* d-*hypoelliptic if, for every open set* $\Omega \subset \mathbb{R}^n$ *and every distribution* $T \in \mathcal{D}'(\Omega)$, $PT \in \Gamma^d(\Omega)$ *implies* $T \in \Gamma^d(\Omega)$.

Condition (dH) above is a necessary one in order that a differential operator be d-hypoelliptic. More precisely, we have

Theorem 1.6. *Let* $P(D)$ *be a partial differential operator. If, for some open set* $\Omega \subset \mathbb{R}^n$, *every distribution* $u \in \mathcal{D}'(\Omega)$ *such that* $Pu = 0$ *belongs to* $\Gamma^d(\Omega)$, *then condition* $(dH)_1$ *holds.*

*Proof.* As in the proof of Theorem 1.2, let

$$H_p(\Omega) = \{u \in L^\infty_{loc}(\Omega) : P(D)u = 0\}$$

Equipped with the topology induced by $L^\infty_{loc}(\Omega)$, $H_p(\Omega)$ is a Frechet space. Define on $H_p(\Omega)$ a second locally convex topology by the family of seminorms

$$S_{k,\nu}(u) = \sup_{x \in K} \sum_p \left(\frac{1}{|p|!}\right)^{d+1/\nu} |D^p u(x)|$$

where K runs over the compact subsets of $\Omega$ and $\nu = 1,2,\ldots$ . Clearly, these seminorms are finite because, by hypothesis, if $u \in H_p(\Omega)$ then $u \in \Gamma^d(\Omega)$. Equipped with the topology defined by the seminorms $(S_{k,\nu})$, $H_p(\Omega)$ becomes a Frechet space. Since these two topologies on $H_p(\Omega)$ are comparable, they coincide. Therefore, if we fix a compact subset $K \subset \Omega$, there is, for every $\nu = 1,2,\ldots$ , a constant $B_\nu$ and a compact subset $H_\nu \subset \Omega$ such that, for all $u \in H_p(\Omega)$, we have

$$S_{k,\nu}(u) \leq B_\nu \, ||u||_{L^\infty(H_\nu)}$$

Let, now, $u = e^{i\langle x,\zeta\rangle}$ with $P(\zeta) = 0$ and set $\delta = \sup_{x \in K} |x|$ and $\delta' = \sup_{x \in H_\nu} |x|$. We obtain

$$\sum_p \left(\frac{1}{|p|!}\right)^{d+1/\nu} |\zeta^p| \leq B_\nu \, e^{(\delta+\delta')|\mathrm{Im}\zeta|}$$

Using the inequality

$$|\zeta|^k \leq n^{k/2} \sum_{|p|=k} |\zeta^p|$$

we derive

$$\sum_{k=0}^{\infty} \left(\frac{1}{k!}\right)^{d+1/\nu} n^{-k/2} |\zeta|^k \leq B_\nu \, e^{(\delta+\delta')|\mathrm{Im}\zeta|}$$

At this point we use the following result

Lemma 1.2. *Let*

$$E^c(t) = \sum_{k=0}^{\infty} \left(\frac{t^k}{k!}\right)^c$$

*There are constants* M, $\gamma > 0$ *such that*

$$\exp(\gamma t) \leq M\, E^c(t), \qquad \forall\, t \geq 0$$

By setting $c = d+1/\nu$ and $t = (|\zeta| n^{-1/2})^{1/c}$ and applying Lemma 1.2, the last inequality implies

$$\exp(\gamma_\nu |\zeta|^{1/(d+1/\nu)}) \leq B'_\nu \, e^{A_\nu |\mathrm{Im}\zeta|}$$

with suitable constants $\gamma_\nu$, $B'_\nu$ and $A_\nu$. This implies that $((d+1/\nu)H)_1$ holds for $\nu = 1,2,\ldots$ . Since, by Theorem 1.5 the set of numbers d' for which $(d'H)_1$ holds form a closed half line $[d_0,\infty)$ with $d_0$ a rational number $\geq 1$, we get, taking limits, $d \geq d_0$, hence $(dH)_1$ holds. Q.E.D.

As we shall see in Section 1.3, condition (dH) is also a sufficient one for d-hypoellipticity. Hence, the following result holds true.

Theorem 1.7. *A differential operator* P(D) *with constant coefficients is* d-*hypoelliptic if, and only if, in some open subset* $\Omega \subset \mathbb{R}^n$, *all distributions solutions of the homogeneous equation* $P(D)u = 0$ *belong to* $\Gamma^d(\Omega)$.

We remark that once it is proved that conditions $(H)_1$, $(H)_2$, $(H)_3$, $(dH)_1$, $(dH)_2$ and $(dH)_3$ are sufficient ones for hypoellipticity, it will follow that every d-hypoelliptic operator is hypoelliptic. Conversely, every hypoelliptic operator is d-hypoelliptic for some $d \geq 1$.

## 1.3 SUFFICIENT CONDITIONS FOR HYPOELLIPTICITY AND d-HYPOELLIPTICITY

We start with the definitions of *fundamental solutions* and *parametrices* of a partial differential operator with constant coefficients in $\mathbb{R}^n$.

Definition 1.4. *We say that a distribution* $E$ *is a fundamental solution of the operator* $P(D)$ *if*

$$P(D)E = \delta \tag{1.5}$$

*where* $\delta$ *denotes the Dirac measure.*

The above formula means that, for every test function $\phi \in C_c^\infty(\mathbb{R}^n)$,

$$\langle P(D)E,\phi\rangle = \langle E,{}^tP(D)\phi\rangle = \phi(0) \tag{1.6}$$

Here, ${}^tP(D)$ denotes the *transpose* of $P(D)$ defined by

$${}^tP(D) = \sum_p (-1)^{|p|} a_p D^p$$

Notice that we can also write ${}^tP(D) = P(-D)$.

Definition 1.5. *A distribution* $E \in \mathcal{D}'(\mathbb{R}^n)$ *is said to be a parametrix of* $P(D)$ *if the distribution*

$$R = P(D)E-\delta$$

*is an integrable function in some open neighborhood of the origin in* $\mathbb{R}^n$. *The distribution* $R$ *is called the rest of the parametrix.*

Hypoelliptic operators can be characterized in terms of regularity properties of their fundamental solutions. We have the following result.

Theorem 1.8. *Let* $P = P(D)$ *be a partial differential operator with constant coefficients. If* $P$ *is hypoelliptic then every fundamental solution is* $C^\infty$ *in* $\mathbb{R}^n\setminus\{0\}$. *Conversely, if there is a fundamental solution which is a* $C^\infty$ *function in* $\mathbb{R}^n\setminus\{0\}$ *then* $P$ *is hypoelliptic.*

*Proof.* The first part follows immediately from the hypoellipticity of $P$ and the fact that $\delta$ is identically zero in $\mathbb{R}^n\setminus\{0\}$. The second part is a consequence of a general result concerning convolution of distributions, which we quote without proof: *if* $E \in \mathcal{D}'(\mathbb{R}^n)$ *is a* $C^\infty$ *function in* $\mathbb{R}^n \setminus \{0\}$ *and* $T \in \mathcal{E}'(\mathbb{R}^n)$, *then* $E_* T$ *is a* $C^\infty$ *function on every open set of* $\mathbb{R}^n$ *where* $T$ *is* $C^\infty$ [8,15].

Assume that $PE = \delta$ and that $E$ is $C^\infty$ in $\mathbb{R}^n \setminus \{0\}$. Let $u$ be a solution of $Pu = f$, with $f$ a $C^\infty$ function in an open set $\Omega \subset \mathbb{R}^n$. We want to prove that $u \in C^\infty(\Omega)$. Since the result to be proved is a local one, we may assume, without loss of generality, that $f$ has compact support. By writing

$$u = \delta_* u = PE_* u = E_* Pu = E_* f$$

it follows, by the result quoted above, that $u \in C^\infty(\Omega)$. Q.E.D.

A well known theorem proved independently by Malgrange [13] and Ehrenpreis [9] (see also, [12] and [16] ) states that *every partial differential operator with constant coefficients possesses a fundamental solution*. This result combined with Theorem 1.8 imply that, in order to show that an operator $P(D)$ is hypoelliptic, it suffices to show that it has *at least* one fundamental solution which is $C^\infty$ in $\mathbb{R}^n \setminus \{0\}$. When this is the case, *all* fundamental solutions will be $C^\infty$ in $\mathbb{R}^n \setminus \{0\}$.

Remark. The following variation of Theorem 1.8 holds true: *If* $P(D)$ *is* d-*hypoelliptic then every fundamental solution belongs to* $\Gamma^d(\mathbb{R}^n \setminus \{0\})$ *and, conversely, if there is a fundamental solution belonging to* $\Gamma^d(\mathbb{R}^n \setminus \{0\})$ *then* $P(D)$ *is* d-*hypoelliptic*.

A proof of this result in the analytic case (i.e., when $d = 1$) can be found in [15].

In view of Theorem 1.8, in order to prove that a partial differential operator $P(D)$ is hypoelliptic, it suffices to construct a fundamental solution which is $C^\infty$ in $\mathbb{R}^n \setminus \{0\}$. Actually, it suffices to construct a parametrix with a smooth rest.

Theorem 1.9. *If a differential operator* $P(D)$, *with constant coefficients, has a parametrix which is a* $C^\infty$ *function in* $\mathbb{R}^n \setminus \{0\}$ *and a rest which is a* $C^\infty$ *function in* $\mathbb{R}^n$, *then* $P(D)$ *is hypoelliptic*.

*Proof*. Let $F \in \mathcal{D}'(\mathbb{R}^n)$ be such that

$$PF = \delta + R$$

with $F \in C^\infty(\mathbb{R}^n \setminus \{0\})$ and $R \in C^\infty(\mathbb{R}^n)$. Let u be a distribution solution of $Pu = f$ and assume that f is $C^\infty$ in some open set $\Omega \subset \mathbb{R}^n$. Write

$$u = \delta_* u = PF_* u - R_* u$$

$$= F_* f - R_* u.$$

Since R is a C function, it follows that $R_* u$ is $C^\infty$ everywhere. Since $F \in C^\infty(\mathbb{R}^n \setminus \{0\})$, by using the result quoted in the proof of Theorem 1.8, it follows that $F_* f$ is $C^\infty$ on every open set where f is $C^\infty$. Q.E.D.

Remark. In general the two convolutions $F_* f$ and $R_* u$ are not well defined. But such a difficulty can be easily overcome by introducing suitable cut-off functions. For more detail, consult [16].

Suppose now, that $P = P(D)$ is a partial differential operator with constant coefficients and satisfying condition (H) of Theorem 1.1. We are going to show that $P(D)$ is a hypoelliptic operator. In view of Theorem 1.9, it suffices to show that $P(D)$ possesses a parametrix belonging to $C^\infty(\mathbb{R}^n \setminus \{0\})$ and the rest belonging to $C^\infty(\mathbb{R}^n)$. First, we observe that condition (H) implies that *there is a number* $R \geq 1$ *such that* $|P(-\xi)| \geq 1$, *for all* $|\xi| \geq R$. Next, let $\chi \in C^\infty(\mathbb{R}^n)$ be such that $\chi(\xi) = 0$, when $|\xi| \leq R$, and $\chi(\xi) = 1$, when $|\xi| \geq R + 1$. Clearly $\chi(\xi)/P(\xi) \in S'(\mathbb{R}^n)$, hence its Fourier transform $E = \mathcal{F}(\chi(\xi)/P(\xi))$ is well defined and belongs to $\mathcal{D}'(\mathbb{R}^n)$. We have,

$$<P(D)E,\phi> = <E,P(-D)\phi> = \int_{\mathbb{R}^n} \chi(\xi)/P(-\xi)\ P(-\xi)\hat{\phi}(\xi)\,d\xi \quad \text{for all } \phi \in C_c^\infty(\mathbb{R}^n)$$

If we set $\psi(\xi) = 1-\chi(\xi)$, we get:

$$<P(D)E,\phi> = \phi(0) - \int_{\mathbb{R}^n} \psi(\xi)\hat{\phi}(\xi)\,d\xi = <\delta,\phi> + <h,\phi>, \quad \forall\ \phi \in C_c^\infty(\mathbb{R}^n)$$

or, equivalently,

$$P(D)E = \delta + h \tag{1.7}$$

Notice that, in this case, the rest h is an *analytic* function in $\mathbb{R}^n$, since it is the Fourier transform of $\psi(\xi) \in C_c^\infty(\mathbb{R}^n)$. The fact that $E \in C^\infty(\mathbb{R}^n \setminus \{0\})$ is a consequence of the following result.

Lemma 1.3. *Let* E *be a distribution in* $\mathbb{R}^n$ *and suppose that to every* $p \in N^n$, *there is an integer* $k \geq 0$ *such that, for every compact* $K \subset \mathbb{R}^n$ *and all* $\phi \in C_c^\infty(\mathbb{R}^n, K)$, *there is a constant* $C = C(p,k,K)$, *independent of* $\phi$, *such that*

$$\sum_{|q|=k} |\langle x^q D^p E, \phi \rangle| \leq C \, ||\phi||_{L^1} \tag{1.8}$$

*Then,* $E \in C^\infty(\mathbb{R}^n \setminus \{0\})$.

*Proof.* The inequality above implies that $x^q D^p E \in L^\infty_{loc}(\mathbb{R}^n)$ for all $q \in N^n$ such that $|q| = k$. It then follows that $|x|^k D^p E \in L^\infty_{loc}(\mathbb{R}^n)$, hence $D^p E \in L^\infty_{loc}(\mathbb{R}^n \setminus \{0\})$ for all $p \in N^n$, therefore, $E \in C^\infty(\mathbb{R}^n \setminus \{0\})$. Q.E.D.

To complete the proof of the hypoellipticity of P, it suffices to show that the tempered distribution $E = \mathcal{F}(\chi(\xi)/P(-\xi))$ satisfies inequality above. The proof, very technical, will not be given here and can be found in [16]. We shall return to this point in Chapter 2 where a complete proof shall be given for parametrices of a hypoelliptic boundary value problem. As for the distribution E, we can show more, namely, that *to every n-tuple* $p \in N^n$ *corresponds an integer* $k > |p|d + nd$ (*where* d *is the real number of Theorem 1.5) such that, for every compact* $K \subset \mathbb{R}^n$ *and all* $\phi \in C_c^\infty(\mathbb{R}^n, K)$, *there is a constant* $C = C(K)$ *independent of* $\phi$, *such that*

$$\sup_{x \in K} |\langle x^q D^p E, \phi \rangle| \leq 2^{|p|} |q|! \, C^{|q|+1} \tag{1.9}$$

*with* $|q| = k$.

Of course (1.9) implies, by Lemma 1.3, that $E \in C^\infty(\mathbb{R}^n \setminus \{0\})$. Hence, by Theorem 1.9, P(D) is hypoelliptic.

Inequality (1.9) implies a stronger result, namely, that E belongs to $\Gamma^d(\mathbb{R}^n \setminus \{0\})$. Indeed, let us choose the integer k so that

$$|p|d + nd < k \leq |p|d + nd + 1$$

From properties of Euler's gamma function $\Gamma(x)$ we get

$$k! \leq \Gamma(|p|d+nd+1) \leq A^{|p|+1}(|p|!)^d$$

where A is a constant independent of p. Replacing above we obtain

$$\sup_{x \in k} |\langle x^q D^p E, \phi \rangle| \leq C^{|p|+1}(|p|!)^d \tag{1.10}$$

with a suitable constant C independent of p. The same argument used in the proof of Lemma 1.3 shows that $E \in \Gamma^d(\mathbb{R}^n \setminus \{0\})$.

The following is a strengthened version of Theorem 1.9

Theorem 1.10. *Let* $P(D)$ *be a partial differential operator with constant coefficients. Assume that* $P(D)$ *has a parametrix* $E$ *satisfying the following properties:*

i) $E \in \Gamma^d(\mathbb{R}^n \setminus \{0\})$,

ii) *the rest of* $E$ *is an analytic function in* $\mathbb{R}^n$. *Then, the operator* $P(D)$ *is* d-*hypoelliptic*.

We can then summarize our results as follows. If $P(D)$ is a hypoelliptic operator, it satisfies condition (H), *a fortiori*, condition (dH) for some $d \geq 1$ [Theorem 1.5]. Its parametrix $E = \mathcal{F}(\chi(\xi)/P(-\xi))$ belongs to $\Gamma^d(\mathbb{R}^n \setminus \{0\})$ and the rest of E is an analytic function in $\mathbb{R}^n$, therefore $P(D)$ is a d-hypoelliptic operator [Theorem 1.10]. Conversely, if $P(D)$ is d-hypoelliptic, its symbol $P(\zeta)$ satisfies condition (dH), hence condition (H), and $P(D)$ is hypoelliptic [Theorem 1.9].

## 1.4 HYPOELLIPTICITY AND $(d_1,\dots,d_n)$-HYPOELLIPTICITY

When considering solutions of the equation $Pu = 0$, with P hypoelliptic, we are led to study the different degrees of regularity of u with respect to each of the variables $x_1,\dots,x_n$. This brings us to another concept of Gevrey classes.

Definition 1.6. *Let* $\Omega$ *be an open subset of* $\mathbb{R}^n$ *and let* $d = (d_1,\dots,d_n)$ *be an n-tuple of nonnegative real numbers. We denote by* $\Gamma^{(d_1,\dots,d_n)}(\Omega)$ *(or* $\Gamma^d(\Omega)$ *when no confusion is possible) the linear space of all functions* $u \in C^\infty(\Omega)$ *such that, to every compact* $K \subset \Omega$, *there is a constant* $C = C(K,u)$ *such that*

$$\sup_{x\in K} |D^p u(x)| \leq C^{|p|+1}(p_1!)^{d_1}\cdots(p_n!)^{d_n}, \quad \forall\, p \in N^n$$

*A function* $u \in \Gamma^{(d_1,\dots,d_n)}(\Omega)$ *is said to be of Gevrey class* $(d_1,\dots,d_n)$.

If $\bar{d} = \max_{1\leq j\leq n} d_j$, then $\Gamma^{(d_1,\dots,d_n)}(\Omega) \subset \Gamma^{\bar{d}}(\Omega)$. To the Gevrey classes of Definition 1.6 corresponds a definition of $(d_1,\dots,d_n)$-hypoellipticity.

Definition 1.7. *We say that a partial differential operator* $P(D)$ *is* d-*hypoelliptic, with* $d$ *an* $n$-*tuple* $(d_1,\dots,d_n)$, *if, for every open set* $\Omega \subset \mathbb{R}^n$ *and every distribution* $T \in \mathcal{D}'(\Omega)$, $PT \in \Gamma^d(\Omega)$ *implies* $T \in \Gamma^d(\Omega)$.

In what follows we shall use the following notation: if $d = (d_1,\dots,d_n)$ is an n-tuple of positive real numbers, set

$$[\xi]_d = \sum_{j=1}^{n} |\xi_j|^{1/d_j}, \qquad \xi \in \mathbb{R}^n \tag{1.11}$$

Theorem 1.11. *Let* $P(\zeta)$ *be a polynomial with constant coefficients, let* $N$ *be its variety of zeros and let* $d = (d_1,\dots,d_n)$ *be an* $n$-*tuple of nonnegative numbers with* $d_j \geq 1$, $1 \leq j \leq n$. *The following are equivalent conditions*

$(dH)_1$ *there is a constant* $C > 0$ *such that* $[\zeta]_d \leq C(1+|\mathrm{Im}\zeta|)$, $\forall\ \zeta \in N$;

$(dH)_2$ *there is a constant* $C' > 0$ *such that* $[\xi]_d \leq C'(1+d(\xi,N))$, $\forall\ \xi \in \mathbb{R}^n$;

$(dH)_3$ *there is a constant* $C'' > 0$ *such that* $[\xi]_d^{|p|}|P^{(p)}(\xi)| \leq C''(1+P(\xi))$, $\forall\ p \in N^n$, $\forall\ \xi \in \mathbb{R}^n$.

Remark 1. Condition $(dH)_1$ is obviously equivalent to the following one

$(dH)_{1'}$ *there is a constant* $C > 0$ *such that* $[\mathrm{Re}\zeta]_d \leq C(1+|\mathrm{Im}\zeta|)$, $\forall\ \zeta \in N$.

Remark 2. Condition $(dH)_3$ implies that $d_j \geq 1$, $1 \leq j \leq n$. Indeed, $(dH)_3$ implies, since $\deg P(\zeta) \geq 1$, that $P(\xi)$ depends effectively on each variable $\xi_j$, $1 \leq j \leq n$. Let $p_j = 1$ and $p_k = 0$, for $k \neq j$. We can choose $\xi_k$, $k \neq j$, such that $\deg_{\xi_j}(\partial_{\xi_j} P) = \deg_{\xi_j} P-1$. Keeping $\xi_k$, $k \neq j$, fixed and letting $|\xi_j| \to +\infty$ we see that if $(dH)_3$ holds then $d_j$ must be $\geq 1$.

*Proof of Theorem 1.11.* $(dH)_1 \Rightarrow (dH)_2$. Let $\xi \in \mathbb{R}^n$. There is $\zeta \in N$ such that $|\xi-\zeta| = d(\xi,N)$, hence $|\mathrm{Im}\zeta| \leq d(\xi,N)$ and $|\xi_j-\zeta_j| \leq d(\xi,N)$, $1 \leq j \leq n$. Since $d_j \geq 1$, we derive from the second inequality

$$[\xi-\zeta]_d \leq n(1+d(\xi,N)) \tag{1.12}$$

On the other hand, from $|\xi_j| \le |\zeta_j| + |\xi_j-\zeta_j|$ and $d_j \ge 1$, we get $[\xi]_d \le [\zeta]_d + [\xi-\zeta]_d$. From $(dH)_1$ and the above we get $(dH)_2$. Q.E.D.

$(dH)_2 \Rightarrow (dH)_1$. If $\zeta \in N$, we have

$$d(\mathrm{Re}\zeta,N) \le |\mathrm{Re}\zeta-\zeta| = |\mathrm{Im}\zeta|$$

By applying $(dH)_2$ with $\mathrm{Re}\zeta$ replacing $\xi$, we get

$$[\mathrm{Re}\zeta]_d \le C'(1+d(\mathrm{Re}\zeta,N)) \le C'(1+|\mathrm{Im}\zeta|)$$

hence $(dH)_1$, in view of Remark 1. Q.E.D.

$(dH)_2 \Rightarrow (dH)_3$. From Lemma 1.1 we get

$$d(\xi,N)\left|\frac{P^{(p)}(\xi)}{P(\xi)}\right|^{1/|p|} \le C, \quad \forall\ \xi \in \mathbb{R}^n \text{ with } P(\xi) \ne 0$$

Hence it follows $P(\zeta)$ being a hypoelliptic polynomial

$$(1+d(\xi,N))\left|\frac{P^{(p)}(\xi)}{1+P(\xi)}\right|^{1/|p|} \le C, \qquad \forall\ \xi \in \mathbb{R}^n$$

Thus $(dH)_2$ implies that

$$[\xi]_d \le C\left(\frac{1+P(\xi)}{P^{(p)}(\xi)}\right)^{1/|p|}, \qquad \forall\ \xi \in \mathbb{R}^n$$

hence $(dH)_3$. Q.E.D.

$(dH)_3 \Rightarrow (dH)_2$. Suppose that $\xi \in \mathbb{R}^n$ is such that $P(\xi) \ne 0$. Then $(dH)_3$ implies with a suitable constant C

$$[\xi]_d \cdot \left|\frac{P^{(p)}(\xi)}{P(\xi)}\right|^{1/ p} \le C$$

hence,

$$[\xi]_d \sum_{|p|\ne 0} \left|\frac{P^{(p)}(\xi)}{P(\xi)}\right|^{1/ p} \le C$$

with another constant. By Lemma 1.1, we get

$$[\xi]_d\, C_m^{-1}(d(\xi,N))^{-1} \le C,$$

hence

$$[\xi]_d \le C_1\ d(\xi,N), \quad \forall\ \xi \in \mathbb{R}^n \text{ such that } P(\xi) \ne 0$$

If $\xi \in \mathbb{R}^n$ is such that $P(\xi) = 0$, then $(dH)_3$ implies

$$[\xi]_d \leq C_2$$

The last two inequalities yield $(dH)_2$. Q.E.D.

Remark. Condition $(dH)_2$ implies trivially condition $(H)_2$. The converse which is also true is a consequence of Seidenberg-Tarski theorem.

Theorem 1.12. *Let* $P(\zeta)$ *be a hypoelliptic polynomial. There is an n-tuple* $(d_1,\ldots,d_n)$ *such that condition* (dH) *holds. The set of numbers* $d_j$, $1 \leq j \leq n$, *for which condition (dH) holds form a closed half line* $[d_j^0,+\infty)$ *with* $d_j^0$ *a rational number greater* than or equal to 1.

*Proof.* We proceed as in the proof of Theorem 1.5 with $|\xi|$ replaced by $|\xi_j|$. We end up with inequalities of the form

$$|\xi_j|^{1/d_j} \leq C(1+d(\xi,N)), \qquad 1 \leq j \leq n$$

which are equivalent to $(dH)_2$. The same reasonings as in Theorem 1.5 prove the last contention of Theorem 1.12. Q.E.D.

In the present situation, we also have the analogous of Theorems 1.2 and 1.6, namely

Theorem 1.13. *Let* $P(D)$ *be a differential operator with constant coefficients and* $d = (d_1,\ldots,d_n)$ *an n-tuple of numbers with* $d_j \geq 1$. *Suppose that for some open set* $\Omega \subset \mathbb{R}^n$, *every* $u \in \mathcal{D}'(\Omega)$ *such that* $Pu = 0$ *belongs to* $u \in \Gamma^d(\Omega)$. *Then, condition* (dH) *holds.*

*Proof.* We define on $H_p(\Omega)$ two topologies, namely,

i) the topology induced by $L^\infty_{loc}(\Omega)$;

ii) the topology defined by the family of seminorms

$$S_{K,\nu}(u) = \sup_{x\in K} \sum_p \left(\frac{1}{p_1!}\right)^{d_1+1/\nu} \cdots\cdots \left(\frac{1}{p_n!}\right)^{d_n+1/\nu} |D^p u(x)|$$

where $K$ runs over the compact subsets of $\Omega$ and $\nu = 1,2,\ldots$ . With minor modifications of the proof of Theorem 1.6 we get that $(dH)_1$ holds. Q.E.D.

For more details the reader should consult [16].

As a consequence, condition (dH) is necessary for the $(d_1,\ldots,d_n)$-hypoellipticity of an operator $P(D)$. To show that (dH) is a sufficient condition, one uses Theorem 1.10, with d replaced by an n-tuple $(d_1,\ldots,d_n)$. As we have shown before [Sec. 1.3], the distribution $E = \mathcal{F}(\chi(\xi)/P\text{-}(\xi)$ defines a parametrix of $P(D)$ with rest on analytic function in $\mathbb{R}^n$. Furthermore, since (dH) holds we can prove the following result: *to every compact* $K \subset \mathbb{R}^n$ *there is a constant* $C = C(K)$ *such that, for every* $p \in N^n$*, there is an integer* $k \leq \sigma|p| + \tau$ *with* $\sigma$ *and* $\tau$ *independent of* p*, such that, for all* $q \in N^n$ *with* $|q| = k$*, we have*

$$|x^q D^q E(x)| \leq C^{|p|+1}(p_1!)^{d_1}\cdots(p_n!)^{d_n}, \quad \forall\, x \in k \tag{1.13}$$

For the proof of the above, the reader should consult [16]. In Chapter 2, we shall prove a similar result for parametrices of a regular hypoelliptic boundary value problem.

We can summarize our results as follows. If $P(D)$ is a hypoelliptic operator then, by Theorem 1.2, it satisfies condition (H). By Theorem 1.12, there is an n-tuple $(d_1,\ldots,d_n)$ such that $P(D)$ satisfies condition (dH). Hence, by Theorem 1.10 and the above remarks, $P(D)$ is a d-hypoelliptic, by Theorem 1.13, it satisfies (dH) which implies condition (H), hence, by the results of Section 1.3, $P(D)$ is hypoelliptic.

## 1.5 SEMIELLIPTIC OPERATORS

Let $P(D)$ be a partial differential operator in $\mathbb{R}^n$ with constant coefficients and degree $\geq 1$. For some n-tuples $m = (m_1,\ldots,m_n)$ of integers $m_j \geq 1$, we can write

$$P(D) = \sum_{|p:m|\leq 1} a_p D^p \tag{1.14}$$

with

$$|p:m| = \sum_{j=1}^{n} \frac{p_j}{m_j} \tag{1.15}$$

Define the *principal part* of $P(D)$ by

$$P_m(D) = \sum_{|p:m|=1} a_p D^p \tag{1.16}$$

and its *principal symbol* by

$$P_m(\xi) = \sum_{|p:m|=1} a_p \xi^p \tag{1.17}$$

In general, the n-tuple m such that P(D) can be represented by (2.2) is not unique.

Definition 1.8. *We say that a partial differential operator* P(D) *is semielliptic if* $P_m(\xi) \neq 0$, *for all* $\xi \in \mathbb{R}^n \setminus \{0\}$.

The classical example of a semielliptic operator is the heat operator $H = \frac{\partial}{\partial t} - \sum_{j=1}^{n} \frac{\partial^2}{\partial x_j^2}$. As other examples we mention the parabolic and k-parabolic operators [16] and, as we shall see below, the elliptic operators.

Lemma 1.4. *If* P(D) *is semielliptic then the n-tuple* $m = (m_1,\ldots,m_n)$ *is unique. Moreover,* $m_j$ *is precisely the degree of* P(D) *with respect to* $D_j$, $1 \leq j \leq n$.

*Proof.* For every j, $1 \leq j \leq n$, take $p \in N^n$ such that $p_j = m_j$ and $p_i = 0$ if $i \neq j$, and let $e_j$ be the j-th unit vector. By assumption, it follows that $P_m(e_j) \neq 0$, hence $m_j$ is the degree of P with respect to the j-th variable. Q.E.D.

Given the n-tuple $m = (m_1,\ldots,m_n)$, set

$$\bar{m} = \max_{1 \leq j \leq n} m_j \quad \text{and} \quad d_j = \frac{\bar{m}}{m_j} \tag{1.18}$$

The principal symbol of P(D) can be written as

$$P_m(\xi) = \sum_{\langle p,d\rangle=\bar{m}} a_p \xi^p \tag{1.19}$$

with $\langle p,d\rangle = p_1d_1+\ldots+p_nd_n$.

Definition 1.9. *We say that a function* f(x) *in* $\mathbb{R}^n$ *is semihomogeneous of degree* k *with respect to a n-tuple* $(d_1,\ldots,d_n)$ *if*

$$f(t^{d_1}x_1,\ldots,t^{d_n}x_n) = t^k f(x_1,\ldots,x_n), \quad \forall\, t > 0$$

According to this definition, the principal symbol of a semielliptic operator is semihomogeneous of degree $\bar{m}$ with respect to $(d_1,\ldots,d_n)$.

Lemma 1.5. *The following are equivalent conditions:*

a) $P_m(\xi) \neq 0$, $\forall\, \xi \in \mathbb{R}^n \setminus \{0\}$;

b) *there is a constant* $c > 0$ *such that*

$$|P_m(\xi)| \geq c[\xi]_d^{\bar{m}}, \quad \forall\, \xi \in \mathbb{R}^n \tag{1.20}$$

c) *there is a constant* $C > 0$ *such that*

$$1 + [\xi]_d^{\bar{m}} \leq C(1 + |P(\xi)|), \quad \forall\, \xi \in \mathbb{R}^n \tag{1.21}$$

*Proof.* a) => b). Since the set $[\eta]_d = 1$ is compact in $\mathbb{R}^n$, let c be the minimum of $|P_m(\eta)|$ on this set. For every $\xi \in \mathbb{R}^n$, set $t = [\xi]_d$ and $\eta_j = \xi_j/t^{d_j}$. We have, by semihomogeneity,

$$P_m(\xi) = t^{\bar{m}} P_m(\eta)$$

hence

$$|P_m(\xi)| \geq c[\xi]_d^{\bar{m}}$$

It is obvious that b) => a).

b) => c). Write $P_m(\xi) = P(\xi) + (P_m(\xi)-P(\xi))$. A semihomogeneity argument shows that

$$|P_m(\xi)-P(\xi)| \leq C_1([\xi]_d^{\bar{m}-1} + 1) \tag{1.22}$$

We then get

$$c[\xi]_d^{\bar{m}} \leq |P_m(\xi)| \leq |P(\xi)| + C_1([\xi]_d^{\bar{m}-1} + 1)$$

Since

$$C_1[\xi]_d^{\bar{m}-1} + 1) \leq \frac{c}{2}[\xi]_d^{\bar{m}}, \text{ for } [\xi]_d \text{ large} \tag{1.23}$$

we get

$$\frac{c}{2}[\xi]_d^{\bar{m}} \leq |P(\xi)|, \text{ for } [\xi]_d \text{ large}$$

hence (1.21).

c) => b). Write now $P(\xi) = P_m(\xi) + (P(\xi)-P_m(\xi))$. Using (1.22) we get

$$c[\xi]_d^{\bar{m}} \leq |P_m(\xi)| + C_1'([\xi]_d^{\bar{m}-1} + 1)$$

and, by (1.23),

$$\frac{c}{2}[\xi]_d^{\bar{m}} \leq |P_m(\xi)|, \text{ for } [\xi]_d \text{ large}$$

hence (1.20) by semihomogeneity. Q.E.D.

Theorem 1.14. *If* P(D) *is a semielliptic operator, then* P(D) *is* $(d_1,\dots,d_n)$*-hypoelliptic.*

*Proof.* By assumption $P_m(\xi)$ has a positive lower bound on the set $\{\xi \in \mathbb{R}^n: [\xi]_d = 1\}$, hence, $|P_m(\xi)|$ has a positive lower bound in some complex neighborhood of this set. Thus, we can find two positive numbers $\varepsilon$ and c such that, $\xi \in C^n$, $[\zeta]_d = 1$ and $[\mathrm{Im}\,\zeta]_d \le \varepsilon[\mathrm{Re}\,\zeta]_d$ imply $|P_m(\zeta)| \ge c$. Since $P_m$ is semihomogeneous we derive that

$$|P_m(\zeta)| \ge c[\xi]_d^{\bar{m}}, \text{ whenever } [\mathrm{Im}\zeta]_d \le \varepsilon[\mathrm{Re}\zeta]_d \tag{1.24}$$

Combining (1.22) and (1.24) we get

$$|P(\zeta)| \ge c[\zeta]_d^{\bar{m}} - C_1([\zeta]_d^{\bar{m}-1} + 1)$$

whenever $[\mathrm{Im}\zeta]_d \le \varepsilon[\mathrm{Re}\zeta]_d$. It then follows that there is another constant $C_2$ such that, $\xi \in C^n$, $[\zeta]_d \ge C_2$ and $[\mathrm{Im}\zeta]_d \le \varepsilon[\mathrm{Re}\zeta]_d$ imply $P(\zeta) \ne 0$.

Next, let $\bar{d} = \max_{1\le j\le n} d_j$, choose $\varepsilon_1 < 1$ so that $\frac{1}{n}(\frac{1}{\varepsilon_1})^{1/\bar{d}} > \frac{1}{\varepsilon}$, take $C \ge \max(1/\varepsilon_1, C_2)$ and consider the set

$$S = \{\zeta \in C^n: [\mathrm{Re}\zeta]_d > C(1 + |\mathrm{Im}\zeta|)\}$$

If $\zeta \in S$, we have $[\zeta]_d > C_2$ and

$$[\mathrm{Re}\zeta]_d > \frac{1}{\varepsilon_1}|\mathrm{Im}\zeta_j|, \qquad 1 \le j \le n$$

Raising both sides of the last inequality to the power $d_j^{-1}$, taking into account that $d_j \ge 1$, $C \ge 1$ and the choice of $\varepsilon_1$, we get

$$[\mathrm{Re}\zeta]_d > \frac{1}{\varepsilon}[\mathrm{Im}\zeta]_d$$

Hence, for every $\zeta \in S$, $P(\zeta) \ne 0$, which implies condition $(dH)_1$ and therefore P(D) is $(d_1,\dots,d_n)$-hypoelliptic. Q.E.D.

Suppose that the n-tuple $(m_1,\dots,m_n)$ is such that all integers $m_j$ are equal and let us denote by m their common value. In this case, $d_j = 1$, $1 \le j \le n$, and $P_m(\xi)$ is nothing but the homogeneous part of degree m of $P(\xi)$. If condition a) of Lemma 1.5 is satisfied, P(D) is said to be an *elliptic* operator. Elliptic operators form the most important class of semielliptic ones. In view of their importance we shall rephrase their definition.

Let $P(D)$ be a partial differential operator of degree $\leq m$ and write

$$P(D) = \sum_{j=0}^{m} P_{m-j}(D)$$

where $P_{m-j}(D)$ denotes the homogeneous term of degree $m-j$.

Definition 1.10. *We say that the operator* $P(D)$ *is elliptic if*

$$P_m(\xi) \neq 0, \quad \forall\ \xi \in \mathbb{R}^n \setminus \{0\} \tag{1.25}$$

As examples we mention the following ones:

1) The Laplace operator $\Delta = \sum_{j=1}^{n} \frac{\partial^2}{\partial x_j^2}$.

2) The Cauchy-Riemann operator $\frac{\partial}{\partial \bar{z}} = \frac{1}{2}(\frac{\partial}{\partial x} + i\frac{\partial}{\partial y})$.

3) Powers of the Laplace operator $\Delta^k$.

In the case of elliptic operators, the equivalent conditions b) and c) of Lemma 1.5 become:

b') *there is a constant* $c > 0$ *such that*

$$|P_m(\xi)| \geq c|\xi|^m, \quad \forall\ \xi \in \mathbb{R}^n \tag{1.26}$$

and

c') *there is a constant* $C > 0$ *such that*

$$1 + |\xi|^m \leq C(1 + |P(\xi)|), \quad \forall\ \xi \in \mathbb{R}^n \tag{1.27}$$

As a consequence of Theorem 1.14, we have:

*Every elliptic operator is* $(1,\ldots,1)$*-hypoelliptic and therefore, is* 1*-hypoelliptic* (1.28)

Actually, with obvious modifications in the proof of Theorem 1.14, we can prove more directly the following result.

Theorem 1.15. *Let* $P(\zeta)$ *be an elliptic polynomial. There is a constant* $C > 0$ *such that*

$$|\mathrm{Re}\zeta| \leq C(1 + |\mathrm{Im}\zeta|), \quad \forall\ \zeta \in N \tag{1.29}$$

Of course, Theorem 1.15 implies that every elliptic polynomial is d-hypoelliptic with $d = 1$. The converse is also true.

Theorem 1.16. *Suppose that* P *is a hypoelliptic polynomial with* d = 1. *Then* P *is elliptic.*

*Proof.* By assumption, there is a constant C" > 0 such that

$$|\xi|^{|p|}|P^{(p)}(\xi)| \leq C''(1 + |P(\xi)|), \ \forall\ \xi \in \mathbb{R}^n, \ \forall\ p \in N^n$$

[Theorem 1.11]. But, for some $p \in N^n$ with $|p| = n$, we shall have $P^{(p)}(\xi) = \text{cst.}$, hence

$$|\xi|^m \leq C'(1 + |P(\xi)|), \ \forall\ \xi \in \mathbb{R}^n$$

which implies condition (1.27) and therefore P is elliptic. Q.E.D.

In view of Theorems 1.6, 1.7, 1.15 and 1.16 we can state the following

Theorem 1.17. *If* P(D) *is an elliptic operator, all distributions solutions of* P(D)u = f *are real analytic functions on every open set where* f *is real analytic. Conversely, if all solutions of* P(D)u = 0 *are analytic functions, the operator* P(D) *is elliptic.*

# CHAPTER 2
# HYPOELLIPTIC BOUNDARY PROBLEMS

## 2.1 INTRODUCTION

We denote by $(x_1,\ldots,x_n,x_{n+1})$ or by $(x,t)$ with $x = (x_1,\ldots,x_n)$ and $t = x_{n+1}$ a variable element of $\mathbb{R}^{n+1}$ the (n+1)-dimensional Euclidean space. The dual variables will be denoted by $(\xi_1,\ldots,\xi_n,\xi_{n+1})$ or $(\xi,\tau)$ with $\xi = (\xi_1,\ldots,\xi_n)$ and $\tau = \xi_{n+1}$. Consider a hypoelliptic operator of the form

$$P(D,D_t) = D_t^\sigma + \sum_{j=1}^{\sigma-1} a_j(D)D_j^{\sigma-j}, \tag{2.1}$$

where $a_j(D)$ is a constant coefficients differential operator in $D = (D_1,\ldots,D_n)$ and its characteristic polynomial

$$P(\xi,\tau) = \tau^\sigma + \sum_{j=1}^{\sigma-1} a_j(\xi)\tau^{\sigma-j}. \tag{2.2}$$

As we have remarked in Section 1.1, it is an easy consequence of condition $(H)_1$ that all real roots of

$$P(\xi,\tau) = 0 \tag{2.3}$$

belong to a compact subset of $\mathbb{R}^{n+1}$. Thus, we can find $r > 0$ sufficiently large so that $\xi \in \mathbb{R}^n$, $|\xi| > r$ imply that equation (2.3) has no real root $\tau$. Since the roots depend continuously of $\xi$ (because we are assuming that in $P(\xi,\tau)$ the coefficient of the highest power of $\tau$ is independent of $\xi$), in each connected component of the complement in $\mathbb{R}^n$ of the closed ball $\bar{B}(0,r)$, the number of roots with *positive imaginary part* is constant.

Definition 2.1. *We say that a hypoelliptic polynomial* $P(\xi,\tau)$ *of the form (2.2) has a determined type* $\mu$ *if the number of zeros of equation (2.3) with positive imaginary part is equal to* $\mu$, *for all* $\xi \in \mathbb{R}^n$ *with* $|\xi|$ *sufficiently large.*

When $n > 1$, the complement of the unit ball in $\mathbb{R}^n$ has only one connected component. It then follows that every hypoelliptic operator is of a determined type. This may not be the case when $n = 1$.

Examples.

1) The Laplace operator $\Delta$ has type 1.
2) The Heat operator $\partial_t - \Delta$ has also type 1.
3) If $n > 1$, every elliptic operator has even side 2m and type of hypoellipticity m.
4) The Cauchy-Riemann operator $\partial_z = \partial_x + i\partial_y$ is elliptic but not of a determined type.

In this chapter, we are interested in studying regularity up to the boundary of certain boundary value problems. Let $\Omega$ be an open subset of $\mathbb{R}^{n+1}_+ = \{(x,t) \in \mathbb{R}^{n+1} : t > 0\}$ with a plane piece of boundary $\omega$ contained in $\mathbb{R}^n_0 = \{(x,t) \in \mathbb{R}^{n+1} : t = 0\}$. Suppose that $P(D,D_t)$, $Q_1(D,D_t),\ldots,Q_\mu(D,D_t)$ are given differential operators and consider solutions of

$$\begin{cases} P(D,D_t)u = 0 \text{ in } \Omega \\ Q_\nu(D,D_t)u|_\omega = 0, \quad 1 \le \nu \le \mu , \end{cases} \tag{2.4}$$

where $u \in C^k(\Omega \cup \omega)$, with k equal to the maximum degree of P and Q, $1 \le \nu \le k$, and where $Q_\nu(D,D_t)u|_\omega$ denotes the restriction of $Q_\nu(D,D_t)u$ to $\omega$. We want to find *necessary and sufficient conditions in order that every solution of (2.4) belongs to* $C^\infty(\Omega \cup \omega)$ *(resp.* $A(\Omega \cup \omega)$, *resp. a Gevrey class in* $\Omega \cup \omega$*).*

Definition 2.2. *Let* d *be a nonnegative real number (resp.* $d = (d_1,\ldots, d_{n+1})$ *a (n+1)-tuple of nonnegative real numbers). We say that a function* $u \in C^\infty(\Omega \cup \omega)$ *belongs to the Gevrey class* d *in* $\Omega \cup \omega$ *if, given a compact set* $K \subset \Omega \subset \omega$, *there is a constant* $C = C(K,u)$ *such that*

$$|D^p u(x)| \le C^{|p|+1}(|p|!)^d$$

(resp. $|D^p u(x)| \le C^{|p|+1}(p_1!)^{d_1}\ldots(p_{n+1}!)^{d_{n+1}}$), *for all* $p = (p_1,\ldots,p_{n+1})$.

*We denote by* $\Gamma^{d}(\Omega \cup \omega)$ *(resp.* $\Gamma^{(d_1,\ldots,d_{n+1})}(\Omega \cup \omega)$*) the linear space of all functions of Gevrey class* d.

In the introduction to his paper [11], Hormander pointed out that, if we are looking for smooth solutions up to the boundary of (2.4), we must take $P(D,D_t)$ to be hypoelliptic (resp. elliptic) otherwise, we could find a solution $u \notin C^{\infty}$ of $Pu = 0$ in $\Omega$ vanishing in a neighborhood of $\omega$. Also, he indicated that in [11] that when $P(D,D_t)$ is elliptic the correct number of boundary conditions is the type $\mu$ of the operator. From now on, we shall always assume that $P(D,D_t)$ is a hypoelliptic operator of type $\mu$ and of the form (2.1) and that we are given $\mu$ boundary operators $Q_1(D,D_t),\ldots,Q_\mu(D,D_t)$.

Definition 2.3. *With the above notations, we say that* $(P;Q_1,\ldots,Q_\mu)$ *defines a regular hypoelliptic boundary value problem if, for all open subset* $\Omega \subset \mathbb{R}_{+}^{n+1}$ *with a plane piece of boundary* $\omega \subset \mathbb{R}_{0}^{n}$*, all solutions* $u \in C^{k}(\Omega \cup \omega)$ *of the boundary problem*

$$\begin{cases} P(D,D_t)u = f \text{ in} \\ Q_\nu(D,D_t)u\big|_{\omega} = g_\nu, \quad 1 \le \nu \le \mu, \end{cases} \tag{2.5}$$

*belong to* $C^{\infty}(\Omega \cup \omega)$ *whenever* $f \in C^{\infty}(\Omega \cup \omega)$ *and* $g_\nu \in C^{\infty}(\omega)$, $1 \le \nu \le \mu$.

Since $P(D,D_t)$ is a hypoelliptic operator, it is $d = (d_1,\ldots,d_n,d_{n+1})$-hypoelliptic for some (n+1)-tuple d [Theorem 1.12]. This suggests the following

Definition 2.4. *We say that* $(P;Q_1,\ldots,Q_\nu)$ *defines a regular hypoelliptic boundary value problem if, for all open subsets* $\Omega \subset \mathbb{R}_{+}^{n+1}$ *with a plane piece of boundary* $\omega \subset \mathbb{R}_{0}^{n}$*, all solutions* $u \in C^{(k)}(\Omega \cup \omega)$ *of (2.5) belong to* $\Gamma^{d}(\Omega \cup \omega)$ *whenever* $f \in \Gamma^{d}(\Omega \cup \omega)$ *and* $g_\nu \in \Gamma^{(d_1,\ldots,d_n)}(\omega)$, $1 \le \nu \le \mu$.

Regular hypoelliptic problems were first studied by Hormander [11] where he gave an algebraic characterization of such problems. His ideas were further developed in [6] where parametrices and their regularity properties were described and in [1], [2], [7] where d-hypoelliptic boundary value problems were discussed and several characterizations given.

In [11] Hormander defined the characteristic function of a boundary-value problem and characterized regular hypoelliptic problems by the behavior near infinity of the zeros of the characteristic function. His characterization is similar to the algebraic conditions for hypoellipticity. In order to define the characteristic function, we shall first describe some elementary properties of the subset of $C^n$ consisting of all $\zeta$ for which the equation $P(\zeta,\tau) = 0$ has precisely $\mu$ roots with positive imaginary part and none that it is real.

## 2.2 THE SET $A$

Let $P(\zeta,\tau)$ be a hypoelliptic polynomial of the form (2.2) and suppose it to be of type $\mu$. Denote by $A$ *the subset of all* $\zeta \in C^n$ *such that* $P(\zeta,\tau) = 0$ *has precisely* $\mu$ *roots (counting multiplicities) with positive imaginary part and none that is real.* Since the roots are continuous functions of $\zeta$, it follows that $A$ is an *open* set in $C^n$. Since P is hypoelliptic, the set $A$ contains a neighborhood of infinity in $\mathbb{R}^n$.

Examples

1) If $P(\zeta,\tau) = \tau^2 + \sum_{j=1}^{n} \zeta_j^2$, then $A$ is the complement in $C^n$ of $i\mathbb{R}^n$ and the neighborhood of infinity above mentioned is $\mathbb{R}^n \setminus \{0\}$.
2) Let $P = \tau - i\zeta^2$ be a polynomial of two variables. Then,

$$A = \{\zeta \in C : \mathrm{Im}(i\zeta^2) = \xi^2 - \eta^2 > 0\}.$$

We shall now prove some geometric properties of the set $A$.

Theorem 2.1. *Assume that* $P(\zeta,\tau)$ *is a hypoelliptic polynomial of type* $\mu$. *Then, given* $A > 0$, *there is* $B > 0$ *such that* $A$ *contains the set*

$$\{\zeta \in C : |\mathrm{Im}\zeta| \le A,\ |\mathrm{Re}\zeta| \ge B\}. \tag{2.6}$$

*Proof.* We first observe that if $P(\zeta)$ is a hypoelliptic polynomial, then condition $(H)_1$ is trivially equivalent to $(H)_{1'}$. To every $A > 0$ there is $B > 0$ such that

$$|\mathrm{Im}\zeta| \le A \quad \text{and} \quad |\mathrm{Re}\zeta| \ge B$$

imply $P(\zeta) \ne 0$.

If $\tau$ is a real number and $\zeta \in C^n$ satisfies (2.6), we have

$$|\mathrm{Im}(\zeta,\tau)| = |\mathrm{Im}\zeta| \le A \quad \text{and} \quad |\mathrm{Re}(\zeta,\tau)| \ge |\mathrm{Re}\zeta| \ge B$$

hence $P(\zeta,\tau) \neq 0$, that is to say, if $\zeta \in C^n$ satisfies (2.6) the equation $P(\zeta,\tau) = 0$ has *no* real root $\tau$. It then follows that on every connected component of the set (2.6), the number of roots $\tau$ of $P(\zeta,\tau) = 0$ with positive imaginary part is *constant*. Since each connected component contains points $\xi \in \mathbb{R}^n$ with $|\xi|$ sufficiently large and for these the number of roots with positive imaginary part is, by assumption $\mu$, our assertion follows at once. Q.E.D.

Theorem 2.2. *Assume that* $P(\zeta,\tau)$ *is a hypoelliptic polynomial of type* $\mu$. *There is a constant* $C > 0$ *such that A contains the set*

$$\{\zeta \in C^n : |\mathrm{Re}\zeta|^{1/d} \geq C(1 + |\mathrm{Im}\zeta|)\} \tag{2.7}$$

*where* $d \geq 1$.

*Proof.* We know that if $P(\zeta)$ is a hypoelliptic polynomial it is d-hypoelliptic for some $d \geq 1$. By condition $(dH)_{1'}$, there is a constant $C > 0$ such that

$$|\mathrm{Re}\zeta|^{1/d} \geq C(1 + |\mathrm{Im}\zeta|)$$

implies $P(\zeta) \neq 0$. If $\tau \in \mathbb{R}$ and $\zeta \in C^n$ satisfies (2.7) we have

$$|\mathrm{Re}(\zeta,\tau)|^{1/d} \geq |\mathrm{Re}\zeta|^{1/d} \geq C(1 + |\mathrm{Im}\zeta|) = C(1 + |\mathrm{Im}(\zeta,\tau)|)$$

hence $P(\zeta,\tau) \neq 0$. The proof can now be completed with the same reasoning of Theorem 2.1. Q.E.D.

When $P(\zeta,\tau)$ is elliptic we can take, in the previous theorem, $d = 1$ and thus state the following

Corollary. *Assume that* $P(\zeta,\tau)$ *is an elliptic polynomial of type* $\mu$. *There is a constant* $C > 0$ *such that A contains the set*

$$\{\zeta \in C^n : |\mathrm{Re}\zeta| \geq C(1 + |\mathrm{Im}\zeta|)\}. \tag{2.8}$$

## 2.3 THE CHARACTERISTIC FUNCTION OF A BOUNDARY PROBLEM

Let $(f_j(\tau))_{1 \leq j \leq \mu}$ be $\mu$ analytic functions of complex variables and let $k(\tau) = \sum_{j=1}^{\mu} (\tau - \tau_j)$ be a polynomial in $\tau$ with $\mu$ complex roots $\tau_1,\ldots,\tau_\mu$ not necessarily distinct. Define

$$R(k;f_1,\ldots,f_\nu) = \frac{\det\,(f_\nu(\tau_j))_{\substack{1\le\mu\le\mu\\ 1<j<\mu}}}{\prod_{k<j}(\tau_j-\tau_k)} \tag{2.9}$$

when the zeros are distinct and by continuity otherwise. If we assume $\tau_1,\ldots,\tau_\mu$ to be $\mu$ independent variables, $R(k;f_1,\ldots,f_\nu)$ is then a function of such variables.

Theorem 2.3. *$R(k;f_1,\ldots,f_\nu)$ is an analytic function of the complex variables $\tau_1,\ldots,\tau_\mu$.*

*Proof.* Let $f(\tau)$ be an analytic function in some open region $\Omega \subset C$ and suppose that $\tau_1,\ldots,\tau_\mu$ are variable elements in $\Omega$. We shall assume, for a moment, that $\tau_1,\ldots,\tau_\mu$ are distinct. Define the divided differences

$$f(\tau_1,\tau_2) = \frac{f(\tau_1)-f(\tau_2)}{\tau_1-\tau_2} \tag{2.10}$$

and

$$f(\tau_1,\tau_2,\ldots,\tau_n) = \frac{f(\tau_1,\ldots,\tau_{n-1})-f(\tau_2,\ldots,\tau_n)}{\tau_1-\tau_n}, \quad n \le \mu. \tag{2.11}$$

If $C$ is a Jordan curve contained in $\Omega$ and surrounding the points $\tau_1,\ldots,\tau_\mu$, we can write

$$f(\tau_1,\tau_2) = \frac{1}{2\pi i}\int_C \frac{f(z)}{(z-\tau_1)(z-\tau_2)}\,dz \tag{2.10'}$$

and

$$f(\tau_1,\tau_2,\ldots,\tau_n) = \frac{1}{2\pi i}\int_C \frac{f(z)}{(z-\tau_1)(z-\tau_2)\ldots(z-\tau_n)}dz, \quad n \le \mu. \tag{2.11'}$$

These formulas show that $f(\tau_1,\ldots,\tau_n)$ ($n \le \mu$) is an analytic function of all its variables and that the divided differences are also well defined when the points $\tau_1,\ldots,\tau_\mu$ are not all distinct. By using (2.11) it is easily seen that (2.9) can be written as

$$R(k;f_1,\ldots,f_\nu) = \det\,(f_\nu(\tau_1,\ldots,\tau_j))_{\substack{1\le\nu\le\mu\\ 1\le j\le\mu}} \tag{2.9'}$$

It then follows that $R$ is an analytic function of the variables $\tau_1,\ldots,\tau_\mu$. Q.E.D.

By Taylor's formula, (2.10) can be written as

$$f(\tau_1,\tau_2) = \int_0^1 f'(s\tau_1 + (1-s)\tau_2)ds. \tag{2.10''}$$

Observe that the last formula makes sense even if $\tau_1 = \tau_2$. An induction argument shows that we can write (2.11) as follows

$$\begin{aligned} &f(\tau_1,\tau_2,\ldots,\tau_n) = \\ &\int_0^1 t_1^{n-2}\int_0^1 t_2^{n-3}\cdots\int_0^1 f^{(n-1)}(t_{n-1}\{\ldots[t_2(t_1\tau_1 + (1-t_1)\tau_2) + \ldots]\ldots\} \\ &+ (1-t_{n-1})\{\ldots[\ldots + (1-t_2)(t_1\tau_{n-1} + (1-t_n)\tau_n]\ldots\})dt_{n-1}\ldots dt_2dt_1, \end{aligned} \tag{2.11''}$$

where the argument of $f^{(n-1)}$ is a point in the convex hull of the points $\tau_1,\tau_2,\ldots,\tau_\mu$. Since

$$\int_0^1 t_1^{n-2}\int_0^1 t_2^{n-3}\cdots(\int_0^1 dt_{n-1})\ldots dt_2\, dt_1 = \frac{1}{(n-1)!}$$

we derive from (2.11'') the following inequality

$$|f(\tau_1,\ldots,\tau_n)| \leq \frac{1}{(n-1)!} \sup_{x\varepsilon K} |f^{(n-1)}(z)| \tag{2.12}$$

where $K$ denotes the convex hull of the points $\tau_1,\ldots,\tau_\mu$. In all this discussion we are assuming f to be analytic in some neighborhood of K. Applying the inequality (2.12) to (2.9') we obtain

$$|R(k;f_1,\ldots,f_n)| \leq \prod_{\nu=1}^{\mu} (\sum_{j=0}^{\mu-1} \sup_{z\varepsilon K} \frac{|f_\nu^{(j)}(z)|}{j!}) \tag{2.13}$$

where K is the convex hull of $\tau_1,\ldots,\tau_\mu$.

Suppose now that $P(D,D_t)$ is a hypoelliptic operator of type $\mu$ and of the form (2.1) and let $Q_1(D,D_t),\ldots,Q_\mu(D,D_t)$ be given differential operators. For every $\zeta \varepsilon A$ define

$$k_\zeta(\tau) = \prod_{j=1}^{\mu} (\tau-\tau_j(\zeta)) \tag{2.14}$$

where $(\tau_j(\zeta))_{1\leq j\leq\mu}$ are the roots of $P(\zeta,\tau) = 0$ with positive imaginary part. The function

$$C(\zeta) = R(k_\zeta;Q_1,\ldots,Q_\mu) = \frac{\det(Q_\nu(\zeta,\tau_j(\zeta)))_{\substack{1\le\nu\le\mu\\ 1\le j\le\mu}}}{\prod_{k<j}(\tau_j(\zeta)-\tau_k(\zeta))}, \quad \zeta \in A, \tag{2.15}$$

is called the *characteristic function* of the boundary-value problem defined by $(P;Q_1,\ldots,Q_\mu)$.

Theorem 2.4. $C(\zeta)$ *is an analytic function in* $A$.

*Proof.* Suppose, for a moment, that $(\zeta_1,\ldots,\zeta_n, \tau_1,\ldots,\tau_\mu)$ are $n + \mu$ independent variables. Since $P$, $Q_1,\ldots,Q_\mu$ are polynomials it follows that

$$\frac{\det(Q_\nu(\zeta,\tau_j))_{\substack{1\le\nu\le\mu\\ 1\le j\le\mu}}}{\prod_{k<j}(\tau_j-\tau_k)} \tag{2.16}$$

is a polynomial in $(\zeta_1,\ldots,\zeta_n, \tau_1,\ldots,\tau_\mu)$. Moreover, the divided differences (2.11) are symmetric functions of the variables $(\tau_1,\ldots,\tau_\mu)$, $n \le \mu$. It follows that (2.16) is a symmetric polynomial of $(\tau_1,\ldots,\tau_\mu)$, hence, by a theorem of elementary algebra, $C(\zeta)$ is a polynomial in the coefficients of the polynomial $k_\zeta$. But these coefficients are the elementary symmetric functions of the roots $\tau_1(\zeta),\ldots,\tau_\mu(\zeta)$ which, by a classical result, are analytic functions of $\zeta$. Therefore, $C(\zeta)$ is an analytic function in $A$. Q.E.D.

For later use, we now describe some estimates for solutions of ordinary differential equations. The results of Sections 2.4, 2.5, 2.6 and 2.7 are borrowed from Hormander [11].

## 2.4 SOME RESULTS ON THE INITIAL VALUE PROBLEM FOR ORDINARY DIFFERENTIAL EQUATIONS

Let

$$k(D_t) = D_t^\mu + a_{\mu-1}D_t^{\mu-1}+\ldots+a_0, \tag{2.17}$$

where $D_t = \frac{1}{i}\frac{\partial}{\partial t}$, be an ordinary differential operator with constant coefficients belonging to C and of order $\mu$. It is well known that the solutions of $k(D_t)u = 0$ are linear combinations of exponential polynomial solutions, that is, solutions of the form $G(t)e^{it\tau}$ where $\tau$ is a zero of the characteristic polynomial

$$k(\tau) = \tau^{\mu} + a_{\mu-1}\tau^{\mu-1}+\ldots+a_0 \tag{2.18}$$

and G(t) a polynomial of degree less than the order of the zero $\tau$.

Let $(q_\nu(D_t))_{1\le\nu\le\mu}$ be given ordinary differential operators. We want to determine a necessary and sufficient condition in order that the only solution of the initial value problem

$$k(D_t)u = 0, \qquad (q_\nu(D_t)u) = 0, \qquad 1 \le \nu \le \mu, \tag{2.19}$$

be the trivial one. Such a condition being satisfied the initial value problem

$$k(D_t)u = 0, \qquad (q_\nu(D_t)u)(0) = \psi_\nu, \qquad 1 \le \nu \le \mu, \tag{2.20}$$

where $(\psi_\nu)_{1\le\nu\le\mu}$ are given constants, will then have a unique solution.

Denote by $\tau_1,\ldots,\tau_\mu$ the zeros (not necessarily distinct) of $k(\tau) = 0$ and introduce the function

$$R(k;q_1,\ldots,q_\mu) = \frac{\det(q_\nu(\tau_j))_{\substack{1\le\nu\le\mu \\ 1\le j\le\mu}}}{\prod_{k<j}(\tau_j-\tau_k)} \tag{2.21}$$

which can be considered as a rational function of the indeterminates $\tau_1,\ldots,\tau_\mu$. Notice that each factor in the denominator divides the numerator hence R is well defined in the case of coinciding zeros. Since $(q_\nu(\tau))_{1\le\nu\le\mu}$ are polynomials, it follows that R is a symmetric polynomial in the variables $(\tau_1,\ldots,\tau_\mu)$, hence it can be expressed as a polynomial in the coefficients of k.

Theorem 2.5. *A necessary and sufficient condition in order that* (2.19) *possess a nontrivial solution is that*

$$R(k;q_1,\ldots,q_\mu) = 0. \tag{2.22}$$

*Proof.* 1. Suppose that the zeros $\tau_1,\ldots,\tau_\mu$ of the polynomial (2.18) are all distinct. Then the general solution of $k(D_t)u = 0$ is

$$u(t) = \sum_{j=1}^{\mu} c_j e^{i\tau_j t}$$

where $(c_j)_{1\le j\le\mu}$ are constants. If u(t) has to satisfy the initial conditions (2.19) we get the linear system of equations

$$\sum_{j=1}^{\mu} c_j q_\nu(\tau_j) = 0, \qquad 1 \le \nu \le \mu.$$

Therefore, (2.19) has a trivial solution if and only if $\det (q_\nu(\tau_j))_{\substack{1\le\nu\le\mu \\ 1\le j\le\mu}} = 0.$

2. Suppose now that $\tau_1,\ldots,\tau_r$ are all the distinct zeros of $k(\tau) = 0$ with multiplicities $\mu_1,\ldots,\mu_r$, respectively, and such that $\mu = \mu_1+\ldots+\mu_r$. As it is well known, the functions

$$(it)^s e^{i\tau_j t} \qquad 0 \le s \le \mu_j, \qquad 1 \le j \le r,$$

from a basis of the vector space of all solutions of $k(D_t)u = 0$. The general solution of this differential equation is then

$$u(t) = \sum_{j,s} c_{js}(it)^s e^{i\tau_j t}$$

where $c_{js}$ are constants to be determined. By Leibniz's formula we have:

$$q_\nu(D_t)((it)^s e^{i\tau_j t}) = \sum_\alpha \frac{D_t^\alpha((it)^s)}{\alpha!} q_\nu^{(\alpha)}(D_t)(e^{i\tau_j t})$$

hence

$$q_\nu(D_t)((it)^s e^{i\tau_j t})\Big|_{t=0} = q_\nu^{(s)}(\tau_j)$$

By imposing that μ is to satisfy the initial conditions (2.19) we get the linear system of equations

$$\sum c_{js}\, q^{(s)}(\tau_j) = 0, \qquad 0 \le s < \mu_j, \qquad 1 \le j \le r.$$

It follows that (2.19) has a nontrivial solution if and only if the determinant

$$\begin{vmatrix} q_1(\tau_1) & q_1'(\tau_1) & \cdots & q_1^{(\mu_1-1)}(\tau_1) & q_1(\tau_2) & \cdots \\ \cdots & \cdots & \cdots & \cdots & \cdots & \cdots \\ q_\mu(\tau_1) & q_\mu'(\tau_1) & \cdots & q_\mu^{(\mu_1-1)}(\tau_1) & q_1(\tau_2) & \cdots \end{vmatrix} \qquad (2.23)$$

is equal to zero. To complete the proof, it suffices to show that $R(k;q_1,\dots,q_\mu)$ is precisely the determinant (2.23) divided by

$$\prod_{j=1}^{r} \prod_{s<\mu_j} s! \prod_{1\le k<j\le r} (\tau_j-\tau_k)^{\mu_j\mu_k}$$

Q.E.D.

Theorem 2.6. *A necessary and sufficient condition in order that (2.20) possess a unique solution is that* $R(k;q_1,\dots,q_\mu) \neq 0$. *The solution is then given by*

$$u(t) = \sum_{\nu=1}^{\mu} \psi_\nu \frac{R(k;q_1,\dots,q_{\nu-1},e^{i\tau t},q_{\nu+1},\dots,q_\mu)}{R(k;q_1,\dots,q_\mu)}. \tag{2.24}$$

*Proof.* Since the right hand side of (2.24) is a continuous function of $\tau_1,\dots,\tau_\mu$ whenever $R(k;q_1,\dots,q_\mu) \neq 0$, it suffices to prove (2.24) when all the roots $\tau_j$ are distinct. In that case, u is given by

$$u(t) = \sum_{j=1}^{\mu} c_j e^{i\tau_j t}, \tag{2.25}$$

with the constants $c_j$, $1 \le j \le \mu$, satisfying the linear system of equations

$$\sum_{j=1}^{\mu} c_j q_\nu(\tau_j) = \psi_\nu, \qquad 1 \le \nu \le \mu,$$

which has a unique solution if and only if $\det (q_\nu(\tau_j))_{\substack{1\le\nu\le\mu \\ 1\le j\le\mu}} \neq 0$. By solving this system with respect to $c_j$ and replacing in the previous formula we get (2.24). Q.E.D.

## 2.5 A REPRESENTATION FORMULA FOR SOLUTIONS OF AN INITIAL-VALUE PROBLEM

Suppose that

$$p(\tau) = \tau^\sigma + b_{\sigma-1}\tau^{\sigma-1}+\dots+b_o$$

is a polynomial with exactly $\mu$ roots (counting multiplicities) $\tau_1,\dots,\tau_\mu$ with imaginary part > 0 and no real roots. If we set $k(\tau) = \prod_{j=1}^{\mu} (\tau-\tau_j)$ then, according to our assumption all zeros of the polynomial $p(\tau)/k(\tau)$ have negative imaginary part.

Let u(t) be a $C^\infty$ function vanishing for large t and suppose that u(t) is a solution of the initial-value problem

$$p(D_t)u = f, \qquad (q_\nu(D_t)u)(0) = \psi_\nu, \qquad 1 \le \nu \le \mu, \tag{2.26}$$

Assuming that the degree of p is greater than the degree of each $q_\nu$, $1 \le \nu \le \mu$, we want to get a representation formula for u in terms of f and $(q_\nu)_{1\le\nu\le\mu}$.

First of all, let us get a fundamental solution for $p(D_t)$. We write

$$g_0(t) = \frac{1}{2\pi}\int_{-\infty}^{+\infty} \frac{e^{i\tau t}}{p(t)}\, d\tau \tag{2.27}$$

and observe that this integral is absolutely convergent when $\sigma \ge 2$. If $\sigma = 1$, it converges for $t \ne 0$. Clearly, we have $p(D_t)g_0(t-s) = \delta_{(s)}$, the Dirac measure at the point s, and we call $g_0(t-s)$ *a fundamental solution of the differential operator* $p(D_t)$ *with pole at* s. Let us modify it and obtain another fundamental solution g(t,s) satisfying the following conditions:

$$q_\nu(D_t)g(t,s)\Big|_{t=0} = 0, \qquad 1 \le \nu \le \mu, \tag{2.28}$$

The function

$$w(s,t) = \sum_{\nu=1}^{\mu} (q_\nu(D_t)g_0)(-s)h_\nu(t)$$

with

$$h_\nu(t) = \frac{R(k;q_1,\dots,q_{\nu-1},e^{i\tau t},q_{\nu+1},\dots,q_\mu)}{R(k;q_1,\dots,q_\mu)} \tag{2.29}$$

according to Theorem 2.6, satisfies the equation $k(D_t)w = 0$, hence $p(D_t)w = 0$, and takes on initial values $(q_\nu(D_t)g_0)(-s)$, $1 \le \nu \le \mu$. Therefore, the function

$$g(t,s) = g_0(t-s)-w(s,t) \tag{2.30}$$

is a fundamental solution of $p(D_t)$ with pole at s and satisfies (2.28).

If we set

$$u_1(t) = \int_0^\infty g(t,s)f(s)\,ds$$

we clearly have

$$p(D_t)u_1(t) = f(t)$$

and

$$(q_\nu(D_t)u_1)(0) = 0, \qquad 1 \leq \nu \leq \mu.$$

Setting $u_2 = u - u_1$, we get:

$$p(D_t)u_2 = 0, \qquad (q_\nu(D_t)u_2)(0) = \psi_\nu, \qquad 1 \leq \nu \leq \mu.$$

But the first equation can be replaced by $k(D_t)u_2 = 0$, because $u_2$ is bounded when $t > 0$ (since $u$ and $u_1$ are bounded) and all the zeros of $p(\tau)/k(\tau)$ are in the lower half plane. By Theorem 2.6, $u_2$ is given by formula (2.24) and we obtain the following representation formula

$$u(t) = \int_0^\infty g(t,s)f(s)\,ds + \sum_{\nu=1}^{\mu} \psi_\nu h_\nu(t) \tag{2.31}$$

We have the following estimates of the fundamental solution $g_0(t)$ which will be needed later on.

Theorem 2.7. *Suppose that the zeros of* $p(\tau)$ *satisfy the inequalities*

$$|\tau| \leq c_1, \qquad |\mathrm{Im}\tau| \geq 1 + c_0, \qquad c_0 > 0$$

*Then, for* $t \neq 0$, *we have*

$$|g_0^{(j)}(t)| \leq 2^{\sigma+j} c_1^j e^{-c_0|t|} \tag{2.32}$$

*where* $g_0^{(j)} = D_t^j g_0$.

*Proof.* Supposing $\sigma = 1$, we can assume that $p(\tau) = \tau - \alpha$, with $\mathrm{Im}\alpha > 0$. We then have:

$$g_0(t) = \begin{cases} i\, e^{i\alpha t} & \text{for} \quad t > 0 \\ 0 & \text{for} \quad t < 0 \end{cases}$$

Hence

$$g_0^{(j)}(t) = i(i\alpha)^j e^{i\alpha t}$$

and

$$|g_0^{(j)}(t)| \leq |\alpha|^j\, e^{-t\mathrm{Im}\alpha} \leq c_1^j\, e^{-c_0 t}, \qquad \forall\, t > 0$$

hence (2.32) holds.

2. Suppose that $\sigma > 1$ and $j = 0$. Displacing the line of integration in formula (2.27) we can write:

$$g_0(t) = \frac{1}{2\pi}\int_{-\infty}^{+\infty}\frac{e^{i(\tau\pm ic_0)t}}{p(\tau\pm ic_0)}\,d\tau . \tag{2.33}$$

From our assumptions it follows that the zeros $\tau_k$ of $p(\tau\pm ic_0)$ are such that $|\mathrm{Im}\tau_k| \geq 1$, hence

$$\frac{1}{|p(\tau\pm ic_0)|} \leq \frac{1}{|\tau-\tau_1||\tau-\tau_2|} \tag{2.34}$$

because $|\tau-\tau_k| \geq 1$, for all k. On the other hand, the right hand side of the last inequality can be estimated by $\frac{1}{2}\left(\frac{1}{|\tau-\tau_1|^2}+\frac{1}{|\tau-\tau_2|^2}\right)$ and, it is easily seen that

$$\int_{-\infty}^{+\infty}\frac{d\tau}{|\tau-\tau_j|^2} \leq \pi, \qquad j = 1,2 . \tag{2.35}$$

Combining (2.33), (2.34) and (2.35) we get

$$|g_0(t)| \leq \frac{1}{2\pi}\int_{-\infty}^{+\infty}\frac{e^{\mp c_0 t}}{p(\tau\pm ic_0)} \leq \frac{1}{2}\,e^{-c_0|t|}$$

which implies the desired inequality.

3. The general case will be proved by induction, assuming that the result has already been proved for derivatives of $g_0$ of order $< j$, when the differential operator is of order $< \sigma$. Let $\alpha$ be a zero of $p(\tau)$. Then $(D_t-\alpha)g_0 = ig_0'-\alpha g_0$ is a fundamental solution of the operator $p(D_t)/D_t-\alpha$. By the induction assumption we have

$$|D_t^{s+1}g_0-\alpha D_t^s g_0| \leq 2^{\sigma-1+s}\,c_1^s\,e^{-c_0|t|}, \qquad s < j.$$

Multiplying by $\sigma^{j-1-s}$ we get:

$$|\alpha^{j-1-s}D_t^{s+1}g_0-\alpha^{j-s}D_t^s g_0| \leq 2^{\sigma-1+s}\,c_1^{j-1}\,e^{-c_0|t|}, \qquad s < j.$$

Adding for $0 \leq s \leq j-1$, we obtain

$$|D_t^j g_0-\alpha^j g_0| \leq c_1^{j-1}\,e^{-c_0|t|}\sum_0^{j-1} 2^{\sigma-1+s} \leq c_1^{j-1}\,e^{-c_0|t|}\,2^{\sigma-1+j}.$$

Hence,

$$\begin{aligned} |D_t^j g_0| &\leq |\alpha|^j |g_0| + c_1^{j-1} e^{-c_0|t|} 2^{\sigma-1+j} \\ &\leq c_1^j 2^\sigma e^{-c_0|t|} + c_1^{j-1} e^{-c_0|t|} 2^{\sigma-1+j} \\ &\leq (2^\sigma + 2^{\sigma-1+j}) c_1^j e^{-c_0|t|} \\ &\leq 2^{\sigma+j} c_1^j e^{-c_0|t|} \end{aligned}$$

Q.E.D.

## 2.6 ESTIMATES FOR SOLUTIONS OF DIFFERENTIAL EQUATIONS

Let $k(D_t)$ be the differential operator given by (2.17) and denote by $\tau_1,\ldots,\tau_\mu$ the zeros of $k(\tau)$. Let $H_k$ be the linear space of all functions defined on the interval $I = [0,1]$ which are solutions of the homogeneous equation $k(D_t)u = 0$. Define

$$M(\tau_1,\ldots,\tau_\mu) = \sup_{u\varepsilon H_k} |u(1)|/\int_0^1 |u(t)|dt.$$

Lemma 2.1. *The functions* $M(\tau_1,\ldots,\tau_\mu)$ *is continuous with respect to all its arguments.*

*Proof.* We can always find a basis $(\phi_j(t;\tau_1,\ldots,\tau_\mu))_{1\leq j\leq\mu}$ of the linear space $H_k$ depending continuously on all variables. Writing $u(t) = \sum a_j\phi_j(t;\tau_1,\ldots,\tau_\mu)$ we have

$$M(\tau_1,\ldots,\tau_\mu) = \sup_{u\varepsilon H_k} |\sum a_j\phi_j(1;\tau_1,\ldots,\tau_\mu)|/\int_0^1 |\sum a_j\phi_j(t;\tau_1,\ldots,\tau_\mu)|dt$$

Observing that the supremum remains the same if we impose the restriction $\sum_{j=1}^{\mu} |a_j| = 1$, the continuity of M follows because the denominator never vanishes and the supremum is being taken over a compact set. Q.E.D.

Suppose now that all the roots of $k(\tau) = 0$ are simple ones with one of them equal to zero. It then follows that there exists an integer $\nu < \mu$ such that the anulus $\nu < |\tau| < \nu + 1$ contains no zeros of k.

Lemma 2.2. *Under the above assumptions, there is a* $C^\infty$ *function* $\phi(x)$ *with compact support in* $(-\frac{1}{2},0)$ *and such that its Fourier-Laplace transform satisfies the following conditions:*

$$\hat{\phi}(\tau) = \begin{cases} 0 & \text{when } k(\tau) = 0 \text{ and } |\tau| \geq \nu + 1 \\ 1 & \text{when } k(\tau) = 0 \text{ and } |\tau| \leq \nu . \end{cases} \tag{2.36}$$

*Moreover, $\phi$ is bounded by a constant depending on $\mu$ only.*

*Proof.* Set

$$k_1(\tau) = \prod_{|\tau_j|<\nu} (\tau-\tau_j) \qquad \text{and} \qquad k_2(\tau) = \prod_{|\tau_j|\geq\nu+1} (1-\frac{\tau}{\tau_j}).$$

Let $\psi(x)$ be a fixed $C^\infty$ function with compact support in $(-\frac{1}{2},0)$ and such that $\hat{\psi}(\tau) \neq 0$ when $|\tau| \leq \mu$. Write

$$\phi = h(D_t)k_2(D_t)\psi$$

where $h(D_t)$ is a differential operator of order lower than that of $k_1$, to be determined. Taking Fourier-Laplace transforms we get

$$\hat{\phi}(\tau) = h(\tau)k_2(\tau)\hat{\psi}(\tau)$$

hence (2.36) will be fulfilled if and only if

$$h(\tau) = \frac{1}{k_2(\tau)\psi(\tau)} \quad \text{when} \quad k_1(\tau) = 0. \tag{2.37}$$

This condition determines the polynomial $h(\tau)$. Indeed, the function $F(\tau) = 1/k_2(\tau)\hat{\psi}(\tau)$ is analytic when $|\tau| < \nu$. On the other hand, the zeros $\tau_1,\ldots,\tau_r$ of $k_1(\tau)$ have absolute values $< \nu$. We want to find a polynomial $h(\tau)$ such that $h(\tau_j) = F(\tau_j)$, $1 \leq j \leq r$. Write

$$h(\tau) = a_0 + a_1(\tau-\tau_1)+\ldots+a_{r-1}(\tau-\tau_1)\ldots(\tau-\tau_{r-1}).$$

By imposing the required conditions we get that the coefficients are the finite differences

$$a_j = F(\tau_1,\ldots,\tau_j), \qquad 0 \leq j \leq r-1,$$

thus,

$$h(\tau) = \sum_{j=1}^{r} F(\tau_1,\ldots,\tau_j)(\tau-\tau_1)\ldots(\tau-\tau_{j-1}).$$

It remains to show that $\phi(x)$ is bounded by a constant depending on $\mu$ only. According to (2.12) we have:

$$|F(\tau_1,\ldots,\tau_j)| \leq \frac{1}{(j-1)!} \sup_{\tau\in K} |F^{(j-1)}(z)|$$

where K is the convex hull of $\tau_1,\ldots,\tau_r$ which is contained in the ball

$\{\tau \in C: |\tau| < \nu\}$. On the other hand, if $|\tau| < \nu$ and $\tau_j$ is a zero of $k_2(\tau)$ we have

$$|(1-\frac{\tau}{\tau_j})| \geq \frac{1}{\nu} > \frac{1}{\mu}$$

hence $|F(\tau)| \leq C(\mu)$, which imply an estimate for $|F(\tau_1,\ldots,\tau_j)|$ depending only on $\mu$. Next, observe that the zeros $\tau_j$ of $k_2(\tau)$ are such that $|\tau_j| \geq 1$, thus $|k_2(\tau)| \leq \prod_{|\tau_j| \geq \nu+1} |\tau-\tau_1|$. It then follows that $|\hat{\phi}(\tau)|$ can be estimated by a sum of terms of the type

$$C(\mu)|\tau-\tau_1|\ldots|\tau-\tau_{j-1}|\Pi|\tau-\tau_\ell||\hat{\psi}(\tau)|, \qquad 1 \leq j \leq r .$$

Since $\hat{\psi} \in \phi$ we get, via inverse Fourier transform, the required bound for $\phi$. Q.E.D.

Theorem 2.8. *Suppose that all zeros of the characteristic polynomial of* $k(D_t)$ *have nonnegative imaginary parts. There is a constant* $\gamma$, *depending only on* $\mu$, *the degree of* $k(D_t)$, *such that, if* $u(t)$ *is a solution of* $k(D_t)u = 0$, *we have:*

$$|u(a)| \leq \gamma a^{-1} \int_0^a |u(t)|dt, \qquad a > 0 \tag{2.38}$$

*and*

$$\int_0^b |u(t)|dt \leq (\frac{b}{a})^\gamma \int_0^a |u(t)|dt, \qquad 0 \leq a \leq b. \tag{2.39}$$

*Proof.* 1. First let us show that (2.38) implies (2.39). Set

$$I(x) = \int_0^x |u(t)|dt, \quad \text{with} \quad x > 0.$$

From (2.38) we get $I'(x) \leq \gamma x^{-1} I(x)$ which implies

$$\int_a^b \frac{I'(x)}{I(x)} dx \leq \gamma \int_a^b \frac{1}{x} dx,$$

that is,

$$\log \frac{I(b)}{I(a)} \leq \log (\frac{b}{a}),$$

hence,

$$I(b) \leq (\frac{b}{a})^\gamma I(a).$$

2. To prove (2.38) it suffices to assume that $a = 1$, the general case reducing to this one by means of the substitution $t = as$. Moreover, we may assume that $k(0) = 0$. Indeed, if $\tau_0$ is a zero of $k(\tau)$ with the smallest imaginary part, write

$$\tilde{k}(\tau) = k(\tau+\tau_0) \quad \text{and} \quad \tilde{u}(t) = u(t)e^{i(1-t)\tau_0}$$

Then, all the zeros of $\tilde{k}$ have nonnegative imaginary parts, $\tilde{k}(0) = 0$ and $k(D_t)u = 0$ implies $\tilde{k}(D_t)\tilde{u} = 0$. But,

$$|\tilde{u}(1)| = |u(1)| \quad \text{and} \quad \int_0^1 |\tilde{u}(t)|dt \leq \int_0^1 |u(t)|dt$$

thus inequality (2.38) follows from the corresponding one where $u$ is replaced by $\tilde{u}$.

3. In order to prove (2.38) when $a = 1$ and $k(0) = 0$, we shall proceed by induction on the degree $\mu$ of $k(\tau)$. The inequality being trivially true when $\mu = 1$, let us assume it true when the degree of $k$ is smaller than $\mu$ ($\mu > 1$). Our aim is to show that $M(\tau_1,\ldots,\tau_\mu)$ is bounded by a constant depending only on $\mu$ when $\operatorname{Im} \tau_j \geq 0$, $1 \leq j \leq \mu$. From the continuity of $M$ (Lemma 2.1), this follows immediately when the absolute values of all zeros $(\tau_j)_{1\leq j\leq\mu}$ of $k(\tau)$ are less than or equal to $\mu$. Therefore, it only remains to consider the case when some zero has a larger absolute value. It suffices to consider only the case when all the zeros are simple ones with one of them equal to zero. The case of multiple zeros follows by a continuity argument. There exists then an integer $\nu < \mu$ such that the annulus $\nu < |\tau| < \nu + 1$ contains no zeros of $k$. We are thus under the assumptions of Lemma 2.2. If $u$ is a solution of $k(D_t)u = 0$, write $u = u_1 + u_2$ where $u_1$ (resp. $u_2$) is the sum of the exponential terms of $u$ whose absolute values of the exponents are $\leq \nu$ (resp. $\geq \nu + 1$). We shall prove that there is a constant $C$ depending only on $\mu$, such that

$$\int_0^{1/2} |u_i(t)|dt \leq C \int_0^1 |u(t)|dt, \qquad i = 1,2 . \tag{2.40}$$

Assuming that (2.40) holds, let us complete the proof of the theorem. Each $u_i(t)$, $i = 1,2$, is a solution of a differential equation of order $< \mu$ and satisfying the assumptions of the theorem. By our induction assumption, (2.38) holds for $u_i$, $i = 1,2$. From (2.38), (2.39) and (2.40) we derive

$$|u_i(1)| \leq \gamma\int_0^1 |u_i(t)|dt \leq \gamma 2^\gamma\int_0^{1/2} |u_i(t)|dt \leq C'\int_0^1 |u(t)|dt, \quad i = 1,2 ,$$

hence the theorem.

4. It remains to show (2.40). Let $\phi(x)$ be the function given by Lemma 2.2 and set

$$U(t) = (u*\phi)(t) = \int u(t-s)\phi(s)ds$$

We have

$$\int_0^{1/2} |U(t)|dt \leq \max |\phi(s)| \int_0^1 |u(t)|dt \tag{2.41}$$

Now, if $v = e^{i\tau t}$ with $k(\tau) = 0$ and $|\tau| \leq \nu$, we have

$$v*\phi(t) = \int e^{i\tau(t-s)}\phi(s)ds = e^{i\tau t}\hat{\phi}(\tau) = v(t)$$

in view of (2.36). Therefore, $U = u_1$ and (2.41) implies (2.40). With a similar argument we get (2.40) when $i = 2$. Q.E.D.

Corollary. *Under the assumptions of Theorem 2.8, we have*

$$\sup_{0<t<b} |u(t)| \leq \gamma(\frac{b}{a})^{\gamma-1} \sup_{0<t<a} |u(t)|, \qquad 0 < a \leq b, \tag{2.42}$$

*for all solutions* $u(t)$ *of* $k(D_t)u = 0$.

*Proof*. From (2.39) we get

$$\int_a^b |u(t)|dt \leq a(\frac{b}{a})^{\gamma} \sup_{0<t<a} |u(t)|$$

and from (2.38) we derive

$$\gamma^{-1}s|u(s)| \leq \int_0^b |u(t)|dt, \qquad \forall\, 0 < s \leq b.$$

Combining these two inequalities we get

$$\gamma^{-1}b \sup_{0<t<b} |u(t)| \leq a(\frac{b}{a})^{\gamma} \sup_{0<t<a} |u(t)|$$

which implies (2.42). Q.E.D.

## 2.7 NECESSARY CONDITIONS FOR REGULAR HYPOELLIPTIC PROBLEMS

Let $P(D,D_t)$ be a hypoelliptic operator of type $\mu$ of the form (2.1) and let $A$ be the set defined in 2.2. Suppose that we are given $\mu$ differential operators $Q_1(D,D_t),\ldots,Q_\mu(D,D_t)$ and let $C(\zeta)$, $\zeta \in A$, be the characteristic function (2.15).

Theorem 2.9. *Suppose for some open set* $\Omega \subset \mathbb{R}_+^{n+1}$ *with a plane piece of boundary* $\omega \subset \mathbb{R}_0^n$, *all solutions* $u \in C^k(\Omega \cup \omega)$ *of the homogeneous boundary-value problem*

$$\begin{cases} P(D,D_t)u = 0 & \text{in} \quad \Omega \\ Q_\mu(D,D_t)u\big|_\omega = 0 & \text{in} \quad \omega, \qquad 1 \le \nu \le \mu, \end{cases} \tag{2.43}$$

*belong to* $C^\infty(\Omega \cup \omega)$. *Then, the following condition is satisfied*

$$\zeta \in A, \quad C(\zeta) = 0 \quad \text{and} \quad |\zeta| \to +\infty \quad \textit{imply} \ \operatorname{Im}\zeta \to +\infty. \tag{2.44}$$

*Proof.* 1. Since differentiability is a local property, we shall assume, in what follows, that $\Omega$ is a bounded set. Let $\Omega'$ be an open subset of $\mathbb{R}_+^{n+1}$ such that its closure is contained in $\Omega \cup \omega$ but not in $\Omega$. Denote by $U$ the linear space of all $u \in C^k(\Omega \cup \omega)$ such that

$$p(u) = \sum_{|\alpha| \le k} \sup_{(x,t)\in\Omega} |D^\alpha u(x,t)|$$

is finite. Clearly $p(u)$ is a norm and equipped with it $U$ becomes a Banach space. Denote by $V$ the linear space of all $v \in C^{k+1}(\Omega')$ with bounded derivatives up to and included in the order $k + 1$. The norm

$$p'(v) = \sum_{|\alpha| \le k+1} \sup_{(x,t)\in\Omega'} |D^\alpha v(x,t)|$$

turns $V$ into a Banach space. By our assumption, $U \subset C^\infty(\Omega \cup \omega)$. Next, let $v = \rho(u)$ denote the restriction of $u \in U$ to $\Omega'$. It is clear that $v \in V$. On the other hand, $\rho : U \to V$ is a closed linear map hence, by the closed graph theorem, $\rho$ is continuous. Therefore, there is a constant $C$ such that

$$\sum_{|\alpha| \le k+1} \sup_{(x,t)\in\Omega'} |D^\alpha u(x,t)| \le C \sum_{|\alpha| \le k} \sup_{(x,t)\in\Omega} |D^\alpha u(x,t)| \tag{2.45}$$

for all $u \in C^k(\Omega \cup \omega)$ satisfying (2.43).

2. We are going to apply (2.45) to exponential solutions of the boundary-value problem (2.43), i.e., solutions of the form

$$u(x,t) = e^{i\langle x,\zeta\rangle} v(t) \tag{2.46}$$

where $\zeta \in C^n$ and $v(t)$ is a suitable function of a single variable $t$. In order that (2.46) be a solution of (2.43) it is necessary and sufficient that v be a solution of the initial-value problem

$$\begin{cases} P(\zeta,D_t)v(t) = 0 \\ Q_\nu(\zeta,D_t)v(t)\big|_{t=0} = 0, \qquad 1 \le \nu \le \mu. \end{cases} \tag{2.47}$$

By Theorem 2.5, it follows that if $\zeta \in A$ and $C(\zeta) = 0$ there is a nontrivial solution of

$$\begin{cases} k_\zeta(D_t)v = 0 \\ Q_\nu(\zeta,D_t)v\big|_{t=0} = 0 \end{cases} \tag{2.48}$$

with $k_\zeta(\tau)$ defined by (2.14). Since $k_\zeta(\tau)$ is a factor of $P(\zeta,\tau)$, $v(t)$ is a nontrivial solution of (2.47). Since differentiation of an exponential solution with respect to a tangential variable $x_j$, $1 \le j \le n$, corresponds to multiplication by $\zeta_j$, we derive from (2.45) the following inequality

$$\Big(\sum_{j=1}^n |\zeta_j|\Big) \sum_{|\alpha|\le k} \sup_{\Omega'} |D^\alpha u(x,t)| \le C \sum_{|\alpha|\le k} \sup_{\Omega} |D^\alpha u(x,t)| \tag{2.49}$$

for the exponential solution (2.46). Now we can write $D^\alpha u(x,t) = e^{i<x,\zeta>} v_\alpha(t)$, with $v_\alpha$ a solution of $k_\zeta(D_t)v = 0$. If we denote by H the supremum of $|x|$ when $(x,t) \in \Omega$, we have:

$$e^{-H|\mathrm{Im}\zeta|} \le |e^{i<x,\zeta>}| \le e^{H|\mathrm{Im}\zeta|}, \qquad (x,t) \in \Omega.$$

On the other hand, if a is an upper bound for t, when $(x,t) \in \Omega'$, and b an upper bound for t, when $(x,t) \in \Omega$, we get from (2.49)

$$\sum_{j=1}^n |\zeta_j| \sum_{|\alpha|\le k} \sup_{0<t<a} |v_\alpha(t)| \le C\, e^{2H|\mathrm{Im}\zeta|} \sum_{|\alpha|\le k} \sup_{0<t<b} |v_\alpha(t)|. \tag{2.50}$$

By the Corollary of Theorem 2.8 we have that

$$\sup_{0<t<b} |v_\alpha(t)| \le \gamma\Big(\frac{b}{a}\Big)^{\gamma-1} \sup_{0<t<a} |v_\alpha(t)|. \tag{2.51}$$

From (2.50) and (2.51) we derive that

$$\sum_{j=1}^n |\zeta_j| \le C_1\, e^{2H|\mathrm{Im}\zeta|}$$

or, equivalently,

$$|\zeta| \le C_2\, e^{C_3|\mathrm{Im}\zeta|} \quad \text{if } \zeta \in A \text{ and } C(\zeta) = 0,$$

which implies (2.44). Q.E.D.

Remarks. 1) Condition (2.44) is obviously a necessary one in order that $(P;Q_1,\dots,Q_\mu)$ defines a regular hypoelliptic boundary value problem.

2) Condition (2.44) is equivalent to the following one

$$\text{Given } A > 0 \text{ we can find } B > 0 \text{ such that } \zeta \in C^n,\ |\mathrm{Im}\zeta| \leq A,\ |\mathrm{Re}\zeta| \geq B \text{ imply } \zeta \in \mathcal{A} \text{ and } C(\zeta) \neq 0 \tag{2.52}$$

Indeed, (2.52) clearly implies (2.44). Conversely, by Theorem 2.1, given $A > 0$, there is $B' > 0$ such that $\mathcal{A}$ contains the set $\{\zeta \in C^n : |\mathrm{Im}\zeta| \leq A,\ |\mathrm{Re}\zeta| \leq B'\}$. On the other hand, if (2.44) holds, given $A$ we can find $B''$ such that $\zeta \in C^n$, $|\mathrm{Im}\zeta| \leq A$ and $|\mathrm{Re}\zeta| \geq B''$ imply $C(\zeta) \neq 0$. It then suffices to take $B = \max(B', B'')$. Q.E.D.

More significantly, by using the Seidenberg-Tarki theorem we can prove the following

Theorem 2.10. *Condition* (2.44) *is equivalent to condition*

$$\text{There are constants } C > 0 \text{ and } d \geq 1 \text{ such that } \zeta \in C^n,\ |\mathrm{Re}\zeta|^{1/d} \geq C(1+|\mathrm{Im}\zeta|) \text{ imply } \zeta \in \mathcal{A} \text{ and } C(\zeta) \neq 0. \tag{2.53}$$

*Proof.* Since (2.53) clearly implies (2.44), it is enough to prove the converse. Suppose that there is a real number $t_0$ such that, for all $\zeta \in \mathcal{A}$ with $|\mathrm{Re}\zeta| > t_0$ we have $C(\zeta) \neq 0$. By Theorem 2.2, there are constants $M > 0$ and $\gamma \geq 1$ such that $\mathcal{A}$ contains the set $\{\zeta \in C^n : |\mathrm{Re}\zeta| \geq M(1+|\mathrm{Im}\zeta|^\gamma)\}$. Taking $M' = \max(M, t_0)$, we see that the set $\{\zeta \in C^n : |\mathrm{Re}\zeta| \geq M'(1+|\mathrm{Im}\zeta|^\gamma)\}$ is contained in $\mathcal{A}$ and $C(\zeta) \neq 0$ in that set. But this implies (2.53).

Next, suppose that for every positive real number $t$ there is $\zeta \in \mathcal{A}$ with $|\mathrm{Re}\zeta| > t$ such that $C(\zeta) = 0$. Define the following function

$$M(t) = \inf\{|\mathrm{Im}\zeta| : \zeta \in \mathcal{A},\ |\mathrm{Re}\zeta| \geq t \text{ and } C(\zeta) = 0\}.$$

One can see that $M(t)$ is the infimum of all $\lambda$ such that the following system of equations and inequalities hold:

$$|\mathrm{Re}\zeta|^2 \geq t^2,\ |\mathrm{Im}\zeta|^2 \geq \lambda^2,\ \lambda > 0,\ P(\xi,\tau) = \prod_{j=1}^{\sigma} (\tau-\tau_j)$$

$$\mathrm{Im}\,\tau_1 > 0,\dots,\mathrm{Im}\,\tau_\mu > 0,\quad \mathrm{Im}\,\tau_{\mu+1} < 0,\dots,\mathrm{Im}\,\tau_\sigma < 0$$

$$k_\zeta(\tau) = \prod_{j=1}^{\mu} (\tau-\tau_j),\quad R(k_\zeta(\tau);Q_1(\zeta,\tau),\dots,Q_\mu(\zeta,\tau)) = 0.$$

From the Seidenberg-Tarksi theorem it follows that $M(t)$ is a piecewise algebraic function of $t$. If condition (2.44) holds, it implies that $M(t) \to +\infty$ as $t \to +\infty$. By using the Puisseux expansion of $M(t)$ at infinity we get:

$$M(t) = ct^{\varepsilon}(1 + O(1)), \qquad t \sim +\infty,$$

with $\varepsilon > 0$ and $C > 0$. Hence we get $t \leq C_1 |M(t)|^{\frac{1}{\varepsilon}}$ when $t$ is large. Taking $t = |\mathrm{Re}\zeta|$ with $\zeta \in A$ and $C(\zeta) = 0$ we get $|\mathrm{Re}\zeta| \leq C_1 |\mathrm{Im}\zeta|^{1/\varepsilon}$, for $|\mathrm{Re}\zeta|$ large. Hence, for every $\zeta \in A$ with $C(\zeta) = 0$ we have

$$|\mathrm{Re}\zeta| \leq C_2(1+|\mathrm{Im}\zeta|^{1/\varepsilon})$$

with $C_2$ a suitable constant, which implies (2.46). Q.E.D.

## 2.8 SUFFICIENT CONDITIONS FOR REGULAR HYPOELLIPTIC PROBLEMS

Our aim is to prove that (2.53) is a sufficient condition in order that $(P;Q_1,\ldots,Q_\mu)$ defines a regular hypoelliptic boundary problem. With the notations of the previous sections set, for all $1 \leq \nu \leq \mu$,

$$H_\nu(\xi,t) = \frac{R(k_\xi;Q_1(\xi,\tau(\xi)),\ldots,e^{i\tau(\xi)t},\ldots,Q_\mu(\xi,\tau(\xi)))}{C(\xi)} \tag{2.54}$$

with $\xi \in \mathbb{R}^n$ and $t \geq 0$. If (2.53) holds, $H_\nu(\xi,t)$ is well defined for $|\xi| > M$, with $M > 0$ a suitable constant, and is the unique solution of the initial-value problem

$$\begin{cases} k_\xi(D_t)\, H_\nu(\xi,t) = 0 \\ Q_\ell(\xi,D_t)\, H_\nu(\xi,t)\big|_{t=0} = \delta_{\ell,\nu}, \qquad 1 \leq \ell \leq \mu, \end{cases}$$

where $\delta_{\ell,\nu}$ is the Kronecker symbol (Theorem 2.6). Also, if we set

$$G(\xi,t) = G_0(\xi,t) - \sum_{\nu=1}^{\mu} (Q_\nu(\xi,D_t)G_0)(\xi,0)\, H_\nu(\xi,t) \tag{2.55}$$

where

$$G_0(\xi,t) = \frac{1}{2\pi}\int_{-\infty}^{+\infty} \frac{e^{i\tau t}}{P(\xi,\tau)}\, d\tau$$

we see that for $|\xi| > M$, $G(\xi,\tau)$ is a solution of the initial-value problem

$$\begin{aligned} &P(\xi,D_t)G(\xi,t) = \delta_t \\ &Q_\ell(\xi,D_t)G(\xi,t)\big|_{t=0} = 0, \qquad 1 \leq \ell \leq \mu, \end{aligned} \tag{2.56}$$

where $\delta_t$ denotes the Dirac measure with respect to the variable t.

Let $\chi(\xi) \in C_c^\infty(\mathbb{R}^n)$ be such that $\chi(\xi) = 1$, for all $|\xi| \leq M$, and $\chi(\xi) = 0$, for all $|\xi| \geq M + 1$. We shall prove that

$$(1-\chi(\xi))G(\xi,t) \quad \text{and} \quad (1-\chi(\xi))\, H_\nu(\xi,t), \qquad 1 \leq \nu \leq \mu, \tag{2.57}$$

define, for all $t \geq 0$, *tempered distributions in* $\mathbb{R}^n$, *that their inverse Fourier transforms*

$$K(x,t) = F_\xi^{-1}((1-\chi(\xi))G(\xi,t)) \tag{2.58}$$

and

$$K_\nu(x,t) = F_\xi^{-1}((1-\chi(\xi))\, H_\nu(\xi,t)), \qquad 1 \leq \nu \leq \mu, \tag{2.59}$$

*belong to* $C^\infty(\overline{\mathbb{R}_+^{n+1}}\setminus\{0\})$ *and satisfy the following boundary-value problems*

$$\begin{cases} P(D,D_t)K = \delta_x \otimes \delta_t - \beta(x)\otimes\delta_t \ \text{ in } \overline{\mathbb{R}_+^{n+1}} \\ Q_\nu(D,D_t)K_\nu\big|_{\mathbb{R}_0^n} = 0, \qquad 1 \leq \nu \leq \mu, \end{cases} \tag{2.60}$$

and

$$\begin{cases} P(D,D_t)K_\ell = 0 \ \text{ in } \mathbb{R}_+^{n+1} \\ Q_\nu(D,D_t)K_\nu\big|_{\mathbb{R}_0^n} = \delta_{\nu,\ell}(\delta_x - \beta(x)), \qquad 1 \leq \nu \leq \mu, \end{cases} \tag{2.61}$$

*where* $\beta(x) \in \phi(\mathbb{R}^n)$.

Once this is proved, the regularity of all solutions of the problem (2.5) follows at once. Indeed, the following representation formula, for solutions of (2.5), holds

$$u(x,t) = K*f + \sum_{\nu=1}^{\mu} K_\nu *' g_\nu \tag{2.62}$$

where $*'$ denotes that we are taking convolutions with respect to the tangential variables, only. The assumption that K, $(K_\nu)_{1\leq\nu\leq\mu} \in C^\infty(\overline{\mathbb{R}_+^{n+1}} \setminus \{0\})$ imply that $u(x,t) \in C^\infty(\Omega \cup \omega)$.

The fact that K and $(K_\nu)_{1\leq\nu\leq\mu}$ satisfy (2.60) and (2.61), respectively, is a matter of simple verification left to the reader. That the functions (2.57) defined tempered distributions and (2.58), (2.59) belong to $C^\infty(\overline{\mathbb{R}_+^{n+1}} \setminus \{0\})$ is a consequence of the following two lemmas.

Lemma 2.3. *Suppose that condition* (2.53) *holds and let* $D = \{\zeta \in C^n : |Re\zeta|^{1/d} \geq C(1+|Im\zeta|)\}$. *Then, there are constants* A *and* a *such that*

$$\frac{1}{|C(\zeta)|} \leq A\,|\zeta|^a, \qquad \forall\, \zeta \in D. \tag{2.63}$$

*Proof.* If $1/|C(\zeta)|$ is bounded in D there is nothing to prove. If not, let E be the set of points in a real Euclidean space whose entries are: $Re\,\zeta_1,\ldots,Re\,\zeta_n$, $Im\,\zeta_1,\ldots,Im\,\zeta_n$, r, t, $Re\,\tau_1,\ldots,Re\,\tau_\sigma$, $Im\,\tau_1,\ldots,Im\,\tau_\sigma$ satisfying the following set of equations and inequalities:

$$P(\zeta,\tau) = \prod_{j=1}^{\sigma} (\tau-\tau_j), \qquad k_\zeta(\tau) = \prod_{j=1}^{\mu} (\tau-\tau_j)$$

$$Im\,\tau_1 > 0,\ldots,Im\,\tau_\mu > 0, \qquad Im\,\tau_{\mu+1} < 0,\ldots,Im\,\tau_\sigma < 0$$

$$C(\zeta) = R(k_\zeta(\tau),Q_1(\zeta,\tau(\zeta)),\ldots,Q_\mu(\zeta,\tau(\zeta)))$$

$$t > 0, \qquad r > 0, \qquad |Re\zeta|^2 = \frac{1}{r^2}, \qquad |C(\zeta)|^{-2} = t^2$$

$$|Re\zeta|^2 \geq M(1+ Im\,\zeta|^d)^2 .$$

Since d can be assumed to be a rational number, E is a semialgebraic set. Next, define the following function

$$t(r) = \sup_{r|Re\zeta|=1} |1/C(\zeta)|$$

Since $C(\zeta)$, is, by assumption, $\neq 0$ in D, there is a number $r_0 > 0$ such that on the sphere $r|Re\zeta| = 1$, where $0 < r < r_0$, the function $1/|C(\zeta)|$ is strictly positive thus t(r) is well defined. The set E being semialgebraic, it can be shown that t(r) is a piecewise algebraic function. Moreover, as $r \to 0$, $t(r) \to +\infty$, otherwise $1/C(\zeta)$ would be bounded in D. By the Puisseux expansion near r = 0, we get:

$$t(r) = a_0(r^{\frac{1}{q}})^{-k_0} + a_1(r^{\frac{1}{q}})^{-k_1} + \ldots$$

with $k_0,k_1\ldots$ positive and $k_0,k_1\ldots.$ Hence we get

$$t(r)/a_0\, r^{-k_0/q} \to 1 \text{ as } r \to 0$$

therefore

$$t(r) \leq A_1 r^{-k_0/q} \quad \text{for } r \text{ small.}$$

Setting $r = |\mathrm{Re}\zeta|^{-1}$ and noticing that $|1/C(\zeta)| \leq t(r)$ inequality (2.63) follows. Q.E.D.

Now, let us set

$$G^{(j)}(\zeta,t) = D_t^j G(\zeta,t) \text{ and } H_\nu^{(j)}(\zeta,t) = D_t^j H_\nu(\zeta,t) \tag{2.64}$$

where $H_\nu(\zeta,t)$ and $G(\zeta,t)$ are defined by (2.54) and (2.55), respectively, and $\zeta \in D$.

Lemma 2.4. *Under the assumptions of Lemma* 2.3, *the functions* $G^j(\zeta,t)$ *and* $H_\nu^{(j)}(\zeta,t)$, $1 \leq \nu \leq \mu$, *are analytic in* D *and there are constants* A, B *and* C *such that*

$$|G^{(j)}(\zeta,t)| \leq A\,|\zeta|^B e^{-Ct\,|\zeta|^{1/d}}, \quad \zeta \in D,\ t \geq 0 \tag{2.65}$$

$$|H^{(j)}(\zeta,t)| \leq A\,|\zeta|^B e^{-Ct\,|\zeta|^{1/d}}, \quad \zeta \in D,\ t \geq 0. \tag{2.66}$$

The proof will be postponed to Section 2.10, where we shall prove a sharper version of Lemma 2.4. In the same way we leave aside the proof that the distributions $K(x,t)$ and $(K_\nu(x,t))_{1\leq\nu\leq\mu}$ are $C^\infty$ functions in $\mathbb{R}_+^{n+1} \setminus \{0\}$ since, in Section 2.10, we shall prove that indeed they belong to appropriate Gevrey classes.

We can summarize the results of the last two sections in the following.

Theorem 2.11. *If* $(P;Q_1,\ldots,Q_\mu)$ *defines a regular hypoelliptic boundary value problem in some open set* $\Omega \subset \mathbb{R}_+^{n+1}$ *with a plane piece of boundary* $\omega \subset \mathbb{R}_0^n$, *then condition* (2.53) *holds. Conversely, if* (2.53) *holds,* $(P;Q_1,\ldots,Q_\mu)$ *defines a regular hypoelliptic boundary value problem on every open set of* $\mathbb{R}_+^{n+1}$ *with a plane piece of boundary in* $\mathbb{R}_0^n$.

## 2.9 NECESSARY CONDITIONS FOR REGULAR d-HYPOELLIPTIC PROBLEMS

As we have seen [Sections 1.3 and 1.4], every hypoelliptic operator in $\mathbb{R}^{n+1}$ is d-hypoelliptic (resp. $(d_1,\ldots,d_{n+1})$-hypoelliptic) for some $d \geq 1$ (resp. some (n+1)-tuple with $d_j \geq 1$, $1 \geq j \geq n+1$) and d-hypoellipticity (resp. $(d_1,\ldots,d_{n+1})$-hypoellipticity) is linked to Gevrey regularity. We are going to study Gevrey regularity, up to the boundary, of hypoelliptic boundary value problems.

We begin by supposing that $P(D,D_t)$ a partial differential operator given in (2.1) is d-hypoelliptic and of type $\mu$. Suppose also that $Q_1(D,D_t),\ldots,Q\ (D,D_t)$ are given partial differential operators. The following theorem gives a necessary condition in order that $(P;Q_1,\ldots,Q_\mu)$ defines a regular d-hypoelliptic boundary value problem.

Theorem 2.12. *Suppose that for some open set* $\Omega \subset \mathbb{R}_+^{n+1}$ *with a plane piece of boundary* $\omega \subset \mathbb{R}_0^n$ *all solutions of the homogeneous boundary-value problem* (2.43) *belong to* $\Gamma^d(\Omega \cup \omega)$. *Then, there is a constant* $C > 0$ *such that the following condition holds:*

$$\zeta \in C^n,\ |\mathrm{Re}\zeta|^{1/d} \geq C(1+|\mathrm{Im}\zeta|) \text{ implies } \zeta \in A \text{ and } C(\zeta) \neq 0. \tag{2.67}$$

*Proof.* 1. Let

$$H(\Omega \cup \omega) = \{u \in C^k(\Omega \cup \omega) : Pu = 0 \text{ in } \Omega,\ Q_\nu u|_\omega = 0,\ 1 \leq \nu \leq \mu\}$$

and consider on $H(\Omega \cup \omega)$ two topologies. The first one, $T_1$, defined by the seminorms

$$p_{K,k}(u) = \sum_{|p| \leq k} \sup_{x \in K} |D^p u(x)|$$

where K runs over the compact subsets of $\Omega\ \ \omega$; the second one, $T_2$, defined by the seminorms

$$S_{K,\nu}(u) = \sup_{x \in K} \sum_p \left(\frac{1}{|p|!}\right)^{d+\frac{1}{\nu}} |D^p u(x)|,$$

where K runs through the compact subsets of $\Omega \cup \omega$ and $\nu = 1,2,\ldots$. Clearly, $C^k(\Omega \cup \omega)$ equipped with $T_1$ is a Frechet space and since $H(\Omega \cup \omega)$ is a closed subspace of $C^k(\Omega \cup \omega)$ it follows that, equipped with $T_1$, $H(\Omega \cup \omega)$ is a Frechet space. On the other hand, it can also be shown, with an argument similar to that of [Lemma 7.3], that $T_2$ makes $H(\Omega \cup \omega)$ into a Frechet space. The two topologies being comparable it follows that $T_1 = T_2$, by the open mapping theorem. Hence, for a fixed compact set $K \subset \Omega \cup \omega$ there is, for each $\nu = 1,2,\ldots$ a compact subset $H_\nu \cup \Omega$ and a constant $B_\nu > 0$ such that, for all $u \in H(\Omega \cup \omega)$ we have

$$S_{K,\nu}(u) \leq B_\nu\, p_{H_\nu,k}(u) \tag{2.68}$$

2. We now apply this inequality to exponential solutions of the homogeneous boundary-value problem (2.43), i.e., solutions of the form $u(x,t) = e^{i<x,\zeta>}v(t)$ with $v(t)$ satisfying the initial-value problem (2.48). By Theorem 2.5, there is a nontrivial solution $v$ of (2.48) if $\zeta \in A$ and $C(\zeta) = 0$. We apply (2.68) to the exponential solution $u$. Considering on the left hand side of (2.68) only the terms involving multi-indices $p = (p_1,\ldots,p_n,\ell)$ with $\ell \leq k$, we get:

$$\sup_{(x,t)\varepsilon K} \sum_p \left(\frac{1}{|p|!}\right)^{d+\frac{1}{\nu}} |\zeta^{p'}||e^{i<x,\zeta>}| \, |v^{(\ell)}(t) \tag{2.69}$$

$$\leq B_\nu \sum_{|p|\leq k} \sup_{(x,t)\varepsilon H_\nu} |\zeta^{p'}| \, |e^{i<x,\zeta>}| \, |v^{(\ell)}(t)|$$

Since $|p| = |p'| + \ell$ and $(|p'|+\ell)! \leq 2^{|p'|+\ell}|p'|!\ell!$ we can replace the left hand side of (2.69) by

$$\sup_{(x,t)\varepsilon K} \sum_p \left(\frac{1}{2^{|p!|}2^{\ell}|p'|!\ell!}\right)^{d+\frac{1}{\nu}} |\zeta^{p'}| \, |e^{i<x,\zeta>}| \, |v^{(\ell)}(t)|$$

which, in turn, can be replaced by

$$\left(\frac{1}{2^k k!}\right)^{d+1} \sum_{p'} \left(\frac{1}{|p'|!}\right)^{d+\frac{1}{\nu}} \left|\left(\frac{\zeta}{2^{d+1}}\right)^{p'}\right| \sup_{(x,t)\varepsilon K} \sum_{\ell\leq k} |e^{i<x,\zeta>}| \, |v^{(\ell)}(t)|$$

If we set $\delta = \sup_{(x,t)\varepsilon K} |x|$, $\delta_\nu = \sup_{(x,t)\varepsilon H_\nu} |x|$, $a = \sup_{(x,t)\varepsilon K} t$ and $b = \sup_{(x,t)\varepsilon H_\nu} t$ and observe that we may assume $K \; H_\nu$, we derive from (2.69) the following inequality

$$\sum_{p'} \left(\frac{1}{|p'|!}\right)^{d+\frac{1}{\nu}} \left|\left(\frac{\zeta}{2^{d+1}}\right)^{p'}\right| \sum_{\ell\leq k} \sup_{0<t<a} |v^{(\ell)}(t)|$$

$$\leq (2^k k!)^{d+1} B_\nu |\zeta|^k e^{(\delta+\delta_\nu)|\mathrm{Im}\zeta|} \sum_{\ell\leq k} \sup_{0<t<b} |v^{(\ell)}(t)|$$

Since each $v^{(\ell)}$ is a solution of (2.48), from the Corollary of Theorem 2.8 we derive that

$$\sum_{\ell\leq k} \sup_{0<t<b} |v^{(\ell)}(t)| \leq C \sum_{\ell\leq k} \sup_{0<t<a} |v^{(\ell)}(t)|$$

with a suitable constant $C$ depending only on $\mu$. Thus, there is a constant

$B'_\nu > 0$ such that

$$\sum_{p'} \left(\frac{1}{|p'|!}\right)^{d+\frac{1}{\nu}} \left|\left(\frac{\zeta}{2^{d+1}}\right)^{p'}\right| \leq B'_\nu \, |\zeta|^k \, e^{(\delta+\delta_\nu)|\mathrm{Im}\zeta|}$$

which implies, with an obvious change of variable,

$$\sum_{p'} \left(\frac{1}{|p'|!}\right)^{d+\frac{1}{\nu}} |\zeta^{p'}| \leq C_\nu \, |\zeta|^k \, e^{D_\nu|\mathrm{Im}\zeta|} \tag{2.70}$$

with $C_\nu$ and $D_\nu$ suitable constants. Next, observe that

$$|\zeta|^r \leq \sum_{|p'|=r} \left(\frac{r!}{p'!}\right)^{\frac{1}{2}} |\zeta^{p'}| \leq n^{r/2} \sum_{|p'|=r} |\zeta^{p'}|$$

hence, from (2.70) it follows that

$$\sum_{r=0}^{\infty} \left(\frac{1}{r!}\right)^{d+\frac{1}{\nu}} \left(\frac{|\zeta|}{\sqrt{n}}\right)^r \leq C_\nu \, |\zeta|^k \, e^{D_\nu|\mathrm{Im}\zeta|}$$

Using Lemma 7.4 of [16], we get:

$$e^{\gamma_\nu|\zeta|^{\frac{1}{d+1/\nu}}} \leq C_\nu \, |\zeta|^k \, e^{D_\nu|\mathrm{Im}\zeta|}$$

with $\gamma_\nu$, C and D suitable constants. From the last inequality we derive

$$|\zeta|^{1/d+\frac{1}{\nu}} \leq M_\nu \,(1+|\mathrm{Im}\zeta|) + k \,\ln|\zeta|$$

and noticing that $\ln |\zeta|/|\zeta|^s \to 0$ as $|\zeta| \to +\infty$, $\forall s$, we get

$$|\zeta|^{1/d+\frac{1}{\nu}} \leq M'_\nu \,(1+|\mathrm{Im}\zeta|).$$

But, as in Theorem 1.5, the set of numbers $\delta$ for which the last inequality holds (with $\delta$ replacing $d+\frac{1}{\nu}$) is a closed half-line $[d^0,+\infty)$ with $d_0$ a rational number $\geq 1$. Letting $\nu \to +\infty$ we get

$$|\zeta|^{1/d} \leq M(1+|\mathrm{Im}\zeta|), \ \forall \ \zeta \in C \text{ such that } C(\zeta) = 0,$$

which implies (2.67). Q.E.D.

Next, suppose that the operator $P(D,D_t)$ is $(d_1,\ldots,d_n,d_{n+1})$-hypoelliptic. Set $d' = (d_1,\ldots,d_n)$ and define, as in Section 1.4,

$$[\zeta]_{d'} = \sum_{j=1}^{n} |\zeta_j|^{1/d_j}, \qquad \forall\, \zeta \in \mathbb{C}^n.$$

We have the following generalizations of Theorem 2.2.

Theorem 2.13. *Suppose that* $P(\zeta,t)$ *is a* $(d_1,\ldots,d_{n+1})$*-hypoelliptic polynomial of type* $\mu$. *There is a constant* $C > 0$ *such that* (cf. 2.2) *contains the set*

$$D = \{\zeta \in \mathbb{C}^n : [\mathrm{Re}\zeta]_{d'} \geq C(1+|\mathrm{Im}\zeta|).$$

*Proof.* If $\zeta \in D$ and $\tau$ is any *real* number, we have

$$|\mathrm{Re}\zeta_1|^{1/d_1}+\ldots+|\mathrm{Re}\zeta_n|^{1/d_n} + |\tau|^{1/d_{n+1}} \geq C(1+|\mathrm{Im}\zeta|) = C(1+|\mathrm{Im}(\zeta,\tau)|).$$

Since P is $(d_1,\ldots,d_{n+1})$-hypoelliptic, condition $(dH)_1$ of Theorem 1.1 implies that $P(\zeta,\tau) \neq 0$, $\forall\, \zeta \in D$, $\forall\, \tau \in \mathbb{R}$. In other words, for all $\zeta \in D$, equation $P(\zeta,\tau) = 0$ has no real roots. Thus, on every connected component of D, the number of roots with positive imaginary part is constant. Since every connected component of D contains $\xi \in \mathbb{R}^n$ with $|\xi|$ sufficiently large and for such $\xi$ the number of roots with positive imaginary part is $\mu$, the set $A$ contains D. Q.E.D.

In what follows, let d denote the (n+1)-tuple $(d_1,\ldots,d_{n+1})$ while d' denotes the n-tuple $(d_1,\ldots,d_n)$.

The following theorem generalizes Theorem 2.12.

Theorem 2.14. *If for some open set* $\Omega \subset \mathbb{R}^{n+1}_+$ *with a plane piece of boundary* $\omega \subset \mathbb{R}^n_0$, *all solutions of the homogeneous problem* (2.43) *belong to* $\Gamma^d(\Omega \cup \omega)$, *then there is a constant* $C > 0$ *such that the following condition holds:*

$$\zeta \in \mathbb{C}^n,\ [\mathrm{Re}\zeta]_{d'} \geq C(1+|\mathrm{Im}\zeta|) \textit{ implies } \zeta \in A \textit{ and } C(\zeta) \neq 0\,. \tag{2.71}$$

*Proof.* The proof is with slight modifications the same as that of Theorem 2.12. The topology $T_1$ remains unchanged while the seminorms defining $T_2$ are now replaced by

$$S_{K,\nu}(u) = \sup_{x\in K} \sum_{p} \left(\frac{1}{p_1!}\right)^{d_1+\frac{1}{\nu}} \cdots \left(\frac{1}{p_{n+1}!}\right)^{d_{n+1}+\frac{1}{\nu}} |D^p u(x)|\,.$$

Applying the corresponding inequality (2.68) to exponential solutions we obtain the following inequality

$$\sup_{(x,t)\varepsilon K} \sum_p \left(\frac{1}{p_1!}\right)^{d_1+\frac{1}{\nu}} \cdots \left(\frac{1}{p_n!}\right)^{d_n+\frac{1}{\nu}} |\zeta^{p'}|\ |e^{i<x,\zeta>}|\ \left(\frac{1}{\ell!}\right)^{d_{n+1}+\frac{1}{\nu}} |v^{(\ell)}(t)|$$

$$\leq B_\nu \sum_{|p|\leq k} \sup_{(x,t)\varepsilon H_\nu} |\zeta^{p'}|\ |e^{i<x,\zeta>}|\ |v^{(\ell)}(t)|$$

With minor changes, which are left to the reader, we derive the following inequality

$$|\zeta_1|^{1/d_1+\frac{1}{\nu}} + \ldots + |\zeta_n|^{1/d_n+\frac{1}{\nu}} \leq M_\nu(1+|\mathrm{Im}\zeta|)$$

whenever $\zeta \in C^n$ is such that $C(\zeta) = 0$. Taking limits, as $\nu \to +\infty$, we obtain (2.71).

Q.E.D.

## 2.10 SUFFICIENT CONDITIONS FOR REGULAR d-HYPOELLIPTIC PROBLEMS

Condition (2.71) is a necessary one in order that $(P;Q_1,\ldots,Q\ )$ define a regular $(d_1,\ldots,d_{n+1})$-hypoelliptic boundary-value problem. We are going to show that (2.71) is also sufficient. To prove it suffices to show that the distributions $K(x,t)$ and $(K_\nu(x,t))_{1\leq\nu\leq\mu}$ and satisfying the boundary-value problems (2.60) and (2.61) belong to $\Gamma^{(d_1,\ldots,d_{n+1})}(\overline{\mathbb{R}^{n+1}_+}\setminus\{0\})$. Once this is done, by using the representation formula (2.62), we obtain, by standard methods, that every solution of (2.5) belongs to $\Gamma^{(d_1,\ldots,d_{n+1})}(\Omega\cup\omega)$. The proof that $K$ and $(K_\nu)_{1\leq\nu<\mu}$ belong to the space $\Gamma^{(d_1,\ldots,d_{n+1})}(\overline{\mathbb{R}^{n+1}_+}\setminus\{0\})$ is based upon several lemmas.

Lemma 2.5. *There are constants* $A$ *and* $a$ *such that*

$$\frac{1}{|C(\zeta)|} \leq A\,|\zeta|^a \tag{2.72}$$

*for all* $\zeta$ *satisfying condition* (2.71).

The proof is essentially that of Lemma 2.3 and, for that reason, is left to the reader. The following lemma is a strengthened version of Lemma 2.4. Recall that $\sigma$ denotes the degree of the operator $P(D,D_t)$ with respect to $D_t$.

Lemma 2.6. *Suppose that condition* (2.71) *holds and let* $D = \{\zeta \in C^n : [\mathrm{Re}\zeta]_{d'} \geq C(1+|\mathrm{Im}\zeta|)\}$. *There are constants* A, B *and* C *such that*

$$|G^{(j)}(\zeta,t)| \leq A^{j+1}|\zeta|^B \exp(-Ct[\zeta]_{d'}) \tag{2.73}$$

*and*

$$|H_\nu^{(j)}(\zeta,t)| \leq A^{j+1}|\zeta|^B \exp(-Ct[\zeta]_{d'}), \qquad 1 \leq \nu < \mu , \tag{2.74}$$

*for all* $\zeta \in D$, *and* $t \geq 0$ *and all* $0 \leq j \leq \sigma-1$.

*Proof.* 1. Since the differential operator P is $(d_1,\ldots,d_{n+1})$-hypoelliptic, there is a constant $C > 0$ such that

$$|\mathrm{Re}\zeta_1|^{1/d_1}+\ldots+|\mathrm{Re}\zeta_n|^{1/d_n}+|\mathrm{Re}\tau|^{1/d_{n+1}} \geq C(1+|\mathrm{Im}(\zeta,\tau)|)$$

implies $P(\zeta,\tau) = 0$. Thus, if $\tau$ is a complex root of $P(\zeta,\tau) = 0$ we must have, with a suitable constant $C_1 > 0$,

$$[\mathrm{Re}\zeta]_{d'} \leq C_1(1+|\mathrm{Im}\zeta|+|\mathrm{Im}\tau|)$$

hence,

$$C_1|\mathrm{Im}\tau| \geq [\mathrm{Re}\zeta]_{d'}-C_1(1+|\mathrm{Im}\zeta|).$$

Assuming, what we always can, that for all $\zeta \in D$

$$[\mathrm{Re}\zeta]_{d'} \geq 2\,C_1(1+|\mathrm{Im}\zeta|),$$

we get

$$C_1|\mathrm{Im}\tau| \geq \frac{1}{2}\,[\mathrm{Re}\zeta]_{d'} \tag{2.75}$$

for all $(\zeta,\tau)$ such that $P(\zeta,\tau) = 0$ and $\zeta \in D$. On the other hand, since $d_j \geq 1$, $|\zeta_j| \leq |\mathrm{Re}\zeta_j| + |\mathrm{Im}\zeta_j|$ implies

$$|\zeta_j|^{1/d_j} \leq |\mathrm{Re}\zeta_j|^{1/d_j} + |\mathrm{Im}\zeta_j|^{1/d_j} \leq |\mathrm{Re}\zeta_j|^{1/d_j} + 1 + |\mathrm{Im}\zeta_j|.$$

It follows that

$$[\zeta]_{d'} \leq [\mathrm{Re}]_{d'} + n(1+|\mathrm{Im}\zeta|) \leq C_2\,[\mathrm{Re}\zeta]_{d'} \tag{2.76}$$

for all $\zeta \in D$. Combining (2.75) and (2.76) we get

$$|\mathrm{Im}\tau| \geq C_3 [\zeta]_{d'}, \tag{2.77}$$

for all $(\zeta,\tau)$ such that $P(\zeta,\tau) = 0$ with $\zeta \in D$.

2. It is easy to see that the roots of $P(\zeta,\tau) = 0$ satisfy the inequality

$$|\tau(\zeta)| \leq A'(|\zeta|^{B'} + 1) \tag{2.78}$$

where A' and B' are suitable constants.

3. From (2.54) we get:

$$H_\nu^{(j)}(\zeta,t) = \frac{R(k_\zeta;Q_1(\zeta,\tau(\zeta)),\ldots,(i\tau(\zeta))^j e^{i\tau(\zeta)t},\ldots,Q_\mu(\zeta,\tau(\zeta)))}{C(\zeta)}$$

On the other hand, we have the estimate

$$|R(k_\zeta;Q_1(\zeta,\tau(\zeta)),\ldots,(i\tau(\zeta))^j e^{i\tau(\zeta)t},\ldots,Q_\mu(\zeta,\tau(\zeta)))| \tag{2.79}$$
$$\leq \prod_{\substack{\ell=1\\ \ell\neq\nu}}^{\mu} \Big(\sum_{k=0}^{\mu-1} \sup_K Q_\ell^{(k)}(\zeta,\tau(\zeta))|\Big)\cdot\Big(\sum_{k=0}^{\mu-1} \sup_K \frac{d^k}{d\tau^k}((i\tau(\zeta))^j e^{i\tau(\zeta)t})|\Big),$$

where K is the convex hull of $\tau_1(\zeta),\ldots,\tau_\mu(\zeta)$ [see (2.13)]. In view of (2.78), the terms $|Q_\ell^{(k)}(\zeta,\tau(\zeta))|$ can be estimated by a power of $|\zeta|$. Next, we have

$$\Big|\frac{d^k}{d\tau^k}((i\tau(\zeta))^j e^{i\tau(\zeta)t})\Big| \leq \sum_{p=0}^{k} \binom{k}{p} \frac{d^p}{d\tau^p}(i\tau(\zeta))^j |t|^{k-p} e^{-t|\mathrm{Im}\zeta|}.$$

From (2.77) and (2.78) and taking into account that when $\zeta \in D$, $[\zeta]_{d'}$ is large, we obtain

$$\Big|\frac{d^k}{d\tau^k}((i\tau(\zeta))^j e^{i\tau(\zeta)t})\Big| \tag{2.80}$$
$$\leq (A')^j\, C_5(|\zeta|^{B'} + 1)^j \exp(-t\, C_4\, [\zeta]_{d'}).$$

Inequalities (2.72), (2.79) and (2.80) imply then the inequality (2.74). With the same reasoning we can prove (2.73). Q.E.D.

Lemma 2.7. *Let* $D = \{\zeta \in \mathbb{C}^n : [\mathrm{Re}\zeta]_{d'} \geq C(1+|\mathrm{Im}\zeta|)\}$. *For all* $\xi \in \mathbb{R}^n$

*such that* $[\xi]_{d'} \geq C + 1$, *there is a constant* M *such that the sphere* $S = \{\zeta \in C^n : |\zeta - \xi| \leq \rho\}$ *with* $\rho = M[\xi]_{d'}$ *is contained in* D. *Moreover, there is a constant* $M_1 > 0$ *such that, for all* $\zeta \in S$, *we have* $|\zeta| \leq M_1|\xi|$.

*Proof.* If $\xi \in \mathbb{R}^n$ is such that $[\xi]_{d'} \geq C + 1$, let $\Delta(\xi)$ be its distance to the boundary $\partial D$ of D. Clearly, there is $c_o > 0$ such that $c_o \leq \Delta(\xi) < +\infty$, for all such $\xi$. Let $\zeta_o \in \partial D$ be such that $\Delta(\xi) = |\zeta_o - \xi|$. From

$$|\xi_j| \leq |\zeta_j^o| + |\xi_j - \zeta_j^o|,$$

taking into account that $\zeta_o$ belongs to the boundary of D, we get:

$$\begin{aligned}
[\xi]_{d'} &\leq [\xi_o]_{d'} + [\xi - \zeta_o]_{d'} \\
&\leq [\mathrm{Re}\ \zeta_o]_{d'} + n(1 + |\mathrm{Im}\ \zeta|) + [\xi - \zeta_o]_{d'} \\
&\leq C'(1 + |\mathrm{Im}\ \zeta_o|) + n(1 + \Delta(\xi)) \\
&\leq C''(1 + \Delta(\xi)) = C''\Delta(\xi)(1 + \frac{1}{\Delta(\xi)}) \\
&\leq C'''\Delta(\xi).
\end{aligned}$$

It suffices to take $M < (C''')^{-1}$ to see that S D. Next, if $\zeta \in S$, we have

$$\begin{aligned}
|\zeta| \leq |\xi| + M[\xi]_{d'} &\leq |\xi|(1+M[\xi]_{d'}/|\xi|) \\
&\leq |\xi|(1+Mn(1 + \frac{1}{|\xi|})) \leq M_1|\xi|
\end{aligned}$$

since $|\xi|$ is bounded below for all $\xi \in \mathbb{R}^n$ such that $[\xi]_{d'} \geq C + 1$.

Q.E.D.

Lemma 2.8. *Let* $F(\zeta)$ *be an analytic function on* D *such that*

$$|F(\zeta)| \leq A|\zeta|^\alpha, \qquad \forall\zeta \in D.$$

*Then, there is a constant* $M > 0$, *depending on* C, $d_1, \ldots, d_n$, *but independent of* A, $\alpha$ *and* $F(\zeta)$ *such that*

$$|D_\xi^p(\xi^q F(\zeta))| \leq A\ M^{|p|+|q|+\alpha}|\xi|^{\alpha+|q|}[\xi]_{d'}^{-|p|} \tag{2.81}$$

*for all* $\xi \in \mathbb{R}^n$ *such that* $[\xi]_{d'} \geq C + 1$.

*Proof.* It suffices to prove (2.81) when $q = 0$. From Lemma 2.7, we get

$$|F(\zeta)| \leq A\ M_1^\alpha|\xi|^\alpha, \qquad \forall\zeta \in S$$

In view of Cauchy's integral formula, we obtain

$$|D_\xi^p F(\xi)| \leq AM_1^\alpha |\xi|^\alpha \frac{p!}{\rho^{|p|}} = A \frac{M_1^\alpha}{M^{|p|}} p! |\xi|^\alpha [\xi]_{d'}^{-|p|},$$

hence (2.81). Q.E.D.

We now have at our disposal all the necessary tools to prove the regularity of the distributions K and $(K_\nu)_{1 \leq \nu \leq \mu}$ in $\overline{\mathbb{R}_+^{n+1}} \setminus \{0\}$. We shall consider $(K_\nu)$ only. The proof for K being essentially the same, will be left to the reader.

Theorem 2.15. *The distribution* $K_\nu(x,t)$, $1 \leq \nu \leq \mu$, *belongs to* $\Gamma^{(d_1,\ldots,d_{n+1})}(\overline{\mathbb{R}_+^{n+1}} \setminus \{0\})$.

*Proof.* We have defined $K_\nu(x,t) = F_\xi^{-1}((1-\chi(\xi))H_\nu(\xi,t))$ with $H_\nu(\xi,t)$ given by (2.54) and with $\chi(\xi) \in C_c^\infty(\mathbb{R}^n)$ equal to 1, when $|\xi| \leq N$, and to zero, when $|\xi| \geq N + 1$. The constant N is chosen in such a way that $|\xi| \geq N + 1$ implies $[\xi]_{d'} \geq C + 1$. It suffices to show the following:

*To every compact* $K \subset \overline{\mathbb{R}_+^{n+1}} \setminus \{0\}$ *there corresponds a constant* $C = C(K,K_\nu)$ *such that, for all* $(n+1)$*-tuples* $(p,j) = (p_1,\ldots,p_n,p_{n+1})$ *with* $p = (p_1,\ldots,p_n)$ *and* $j = p_{n+1}$, *we have*

$$\sup_{(x,t)\in K} |D^p D_t^j K_\nu(x,t)| \leq C^{|p|+j+1}(p_1!)^{d_1}\ldots(p_n!)^{d_n}(p_{n+1}!)^{d_{n+1}} \tag{2.82}$$

If $q = (q_1,\ldots,q_n)$ is a n-tuple of nonnegative integers, the integral

$$x^q D^p D_t^j K_\nu(x,t) = (2\pi)^{-n}\int_{\mathbb{R}^n} e^{-i<x,\xi>}[D_\xi^q((1-\chi(\xi))\xi^p H_\nu^{(j)}(\xi,t)]d\xi$$

splits into the sum of the integrals

$$T_o = (2\pi)^{-n}\int_{\mathbb{R}^n} e^{-i<x,\xi>}(1-\chi(\xi))D_\xi^q(\xi^p H_\nu^{(j)}(\xi,t))d\xi$$

and

$$T_{r,s} = (2\pi)^{-n}\int_{\mathbb{R}^n} e^{-i<x,\xi>} D_\xi^r(1-\chi(\xi))D_\xi^s(\xi^p H_\nu^{(j)}(\xi,t))d\xi$$

where $r + s = q$ with $r > 0$. The first integral is estimated as follows. First, we write

$$\frac{1}{q!}D_\xi^q(\xi^p H_\nu^{(j)}(\xi,t)) = \sum_{\substack{q'+q''=q \\ q'\leq q}} \frac{1}{q'!}D^{q'}(\xi^p)\frac{1}{q''!}D^{q''}(H_\nu^{(j)}(\xi,t))$$

Next, observe that

$$\frac{1}{q'!}|D^{q'}(\xi^p)| = \frac{p!}{(p-q')!q'!}|\xi^{p-q'}| \le 2^{|p|}|\xi^{p-q'}|. \tag{2.83}$$

From Lemmas 2.6 and 2.8, we derive

$$|D^{q''}(H_\nu^{(j)}(\xi,t))| \le A^{j+1}C^{|q''|+B}q''!|\xi|^B[\xi]_{d'}^{-|q''|} \quad \forall\ 0 \le j \le \sigma-1. \tag{2.84}$$

Combining (2.83) and (2.84), we get

$$|T_0| \le q!\ 2^{|p|}A^{j+1}\sum_{\substack{q'+q''=q \\ q'\le p}} C^{B+|q''|}\int_{|\xi|\ge N+1}|\xi^{p-q'}||\xi|^B[\xi]_{d'}^{-|q''|}d\xi.$$

Since

$$|\xi|^B \le n^{B/2}\sum_{|r|=B}|\xi^r|,$$

we get

$$|T_0| \le q!\ 2^{|p|}A^{j+1}\sum_{|r|=B}\ \sum_{\substack{q'+q''=q \\ q'\le p}} C^{B+|q''|}\int_{|\xi|\ge N+1}|\xi^{p-q'+r}|[\xi]_{d'}^{-|q''|}d\xi \tag{2.85}$$

At this point, we introduce the following change of variables [10]. Let $d_j \ge 1$, $1 \le j \le n$, and let $x = (x_1,\ldots,x_n) \in \mathbb{R}^n$. There is one and only one solution $\rho(x)$ of $F(x,\rho(x)) = 1$, where

$$F(x,\rho) = \sum_{j=1}^{n}\frac{x_j^2}{\rho^{2d_j}}$$

The element

$$\left(\frac{x_1}{\rho^{d_1}(x)},\ldots,\frac{x_n}{\rho^{d_n}(x)}\right) \in \Sigma_n$$

where $\Sigma_n$ is the unit sphere of $\mathbb{R}^n$. Since

$$\rho(x,y) \le \rho(x) + \rho(y)$$

$\rho$ defines a metric in $\mathbb{R}^n$. For the change of variables

$$x_1 = \rho^{d_1}\cos\Theta_1\ldots\cos\Theta_{n-2}\cos\Theta_{n-1}$$

$$x_2 = \rho^{d_2} \cos \Theta_1 \dots \cos \Theta_{n-2} \sin \Theta_{n-1}$$

$$\vdots$$

$$x_n = \rho^{d_n} \sin \Theta_1$$

we have

$$dx = \rho^{(\sum d_j)-1} J(\Theta_1, \dots, \Theta_n) \, d\rho \, d\Theta$$

with $1 \leq J \leq C$. With such a change of variable, we get

$$\int_{|\xi| \geq N+1} |\xi^{p-q'+r}| [\xi]_d^{-|q''|} d\xi$$

$$\leq \int_{N'}^{\infty} \int_E \rho^{\sum p_j d_j - \sum q_j' d_j + \sum r_j d_j - |q''| + \sum d_j - 1} J(\Theta_1, \dots, \Theta_n) \, d\rho \, d\Theta$$

Hence, all integrals appearing on the right hand side of (2.85) converge provided that

$$\sum p_j d_j - |q| + \sum d_j - 1 + B\bar{d} < -1$$

i.e.

$$|q| > \sum p_j d_j + \sum d_j + B\bar{d}$$

where $\bar{d} = \max_{1 \leq j \leq n} d_j$. Hence, from (2.85), we derive

$$|T_0| \leq |q|! \, C_1^{|p|+|q|+j+1}, \qquad 0 \leq j \leq \sigma-1,$$

with a suitable constant $C_1$. If we also choose $|q| \leq \sum p_j d_j + \sum d_j + B\bar{d} + 1$ and take into account properties of Euler's Gamma function, we get

$$|q|! \leq \Gamma(\sum p_j d_j + \sum d_j + B\bar{d} + 1) \leq A_1^{|p|+1} (p_1!)^{d_1} \dots (p_n!)^{d_n},$$

hence

$$|T_0| < A_1^{|p|+1} C_1^{|p|+|q|+j+1} (p_1!)^{d_1} \dots (p_n!)^{d_n} .$$

Since $|q| \leq |p|\alpha + \beta$, with $\alpha$ and $\beta$ independent of p, we obtain

$$|T_0| \leq A^{|p|+j+1} (p_1!) \dots (p_n!)^{d_n}, \qquad 0 \leq j \leq \sigma-1, \tag{2.86}$$

with A a constant independent of p and q.

The integrals $T_{r,s}$ are estimated as follows. Let $B_r > 0$ be such that

$$|D^r(1-\chi(\xi))| < B_1, \quad \forall\, |r| \le q, \quad \forall\, N < |\xi| < N+1$$

By using (2.83) and (2.84), we get:

$$|T_{r,s}| \le B_1 s! 2^{|p|} A^{j+1} \sum_{s'+s''=s} C^{B+|s''|} \int_{N \le |\xi| \le N+1} |\xi^{p-s'}| |\xi|^B [\xi]_{d'}^{-|s''|} d\xi$$

$$\le B_1 s! 2^{|p|} A^{j+1} C_2^{|q|}$$

$$\le |q|!\, C_3^{|p|+|q|+j+1}, \qquad 0 \le j \le \sigma-1.$$

Proceeding as before we get

$$|T_{r,s}| \le A^{|p|+j+1} (p_1!)^{d_1} \cdots (p_n!)^{d_n}, \qquad 0 \le j \le \sigma-1, \tag{2.87}$$

where A is a constant independent of p and q.

Combining inequalities (2.86) and (2.87), we reach the following conclusion: *To every compact set* $K \subset \overline{\mathbb{R}^{n+1}_+}$ *there is a constant* $C = C(K, K_\nu)$ *such that, for every* $(n+1)$*-tuple* $(p_1, \ldots, p_n, p_{n+1}) = (p, j)$ *with* $0 \le j \le \sigma-1$, *there is a* $n$*-tuple* $q = (q_1, \ldots, q_n)$ *with*

$$|q| \le |p|\alpha + \beta,$$

*where* $\alpha$ *and* $\beta$ *are independent of* p, *such that*

$$\sup_{(x,t)\in K} |x^q D^p D_t^j K_\nu(x,t)| \le C^{|p|+j+1} (p_1!)^{d_1} \cdots (p_n!)^{d_n}, \qquad 0 \le j \le \sigma-1.$$

To obtain (2.82) we make use of $P(D, D_t) K_\nu = 0$. Recalling that $P(D, D_t)$ is given by (2.1) we can write

$$D_t K_\nu = \sum_{j=0}^{\sigma-1} P_j(D) D_t^j K_\nu \tag{2.88}$$

with $P_j(D)$ a differential operator in the tangential variables, with constant coefficients, and order $\le m-j$, where m denotes the total order of $P(D, D_t)$. For simplicity of notations, let us assume that $m = \sigma$, the proof that follows being essentially the same in the general case.

Let us prove, by induction, that to every compact $K \subset \overline{\mathbb{R}^{n+1}_+}$ there is a constant $M \ge 1$ such that

$$\sup_{(x,t)\in K} |x^q D^p D_t^j K_\nu(x,t)| \le C^{|p|+j+1} M^j ((p_1+j)!)^{d_1} \dots ((p_n+j!)^{d_n} (j!)^{d_{n+1}-\sum_{\ell=1}^{n} d_\ell} \tag{2.89}$$

for all p and all j, where C is the constant appearing in the last inequality which we may assume $\ge 1$.

Of course, inequality (2.89) is true when $0 \le j \le \sigma-1$. Suppose it true for $j < J + \sigma$ with $J \ge 0$. Let us show it for $J + \sigma$. Differentiating equation (2.88) we get:

$$D^p D_t^{J+\sigma} K_\nu = \sum_{j=0}^{\sigma-1} P_j(D) D^p D_t^{j+J} K_\nu$$

Let $a_{j\mu} D^\mu$, with $|\mu| < \sigma-j$, be a term of $P_j(D)$. By the induction assumption, we have the following estimate:

$$\begin{aligned} &\sup_{(x,t)\in K} |a_{j\mu} x^q D^{p+\mu} D_t^{j+J} K_\nu(x,t)| \\ &\le |a_{j\mu}| C^{|p|+|\mu|+j+J+1} M^{j+1} ((p_1+\mu_1+j+J)!)^{d_1} \dots ((p_n+\mu_n+j+J)!)^{d_n} \cdot \\ &\qquad ((j+J)!)^{d_{n+1}-\sum_{\ell=1}^{n} d_\ell} \\ &\le |a_{j\mu}| C^{|p|+J+\sigma+1} M^{J+\sigma-1} ((p_1+J+\sigma)!)^{d_1} \dots ((p_n+J+\sigma)!)^{d_n} \cdot \\ &\qquad ((J+\sigma)!)^{d_{n+1}-\sum_{\ell=1}^{n} d_\ell} \end{aligned}$$

Choosing $M > \sum_{j=0}^{\sigma-1} \sum_{|\mu|=0}^{\sigma-1} |a_{j\mu}|$, we obtain

$$\begin{aligned} \sup_{(x,t)\in K} |x^q D^p D_t^{J+\sigma} K_\nu(x,t)| &\le C^{|p|+J+\sigma+1} M^{J+\sigma} \cdot \\ &((p_1+J+\sigma)!)^{d_1} \dots ((p_n+J+\sigma))^{d_n} ((J+\sigma)!)^{d_{n+1}-\sum_{\ell=1}^{n} d_\ell}, \end{aligned}$$

hence (2.89) holds $\forall p$ and $\forall j$.

Finally, using the inequality $(a+b)! \le 2^{a+b} a!b!$ and combining the various constants, it is not difficult to show that (2.89) implies (2.82) with a suitable constant $C = C(K,K_\nu)$ independent of p and q. Arguing as in the proof of Lemma 1.3, we conclude that $K_\nu \in \Gamma^{(d_1,\dots,d_{n+1})}(\overline{\mathbb{R}_+^{n+1}} \setminus \{0\})$.

## 2.11 A PARAMETRIX OF THE BOUNDARY PROBLEM

Let

$$C_c^\infty(\overline{\mathbb{R}_+^{n+1}}; \mathbb{R}_0^n, \mu) = C_c^\infty(\overline{\mathbb{R}_+^{n+1}}) \times C_c^\infty(\mathbb{R}_0^n) \times \dots \times C_c^\infty(\mathbb{R}_0^n)$$

where, in the last product, there are $\mu$ copies of $C_c^\infty(\mathbb{R}_0^n)$. An element of $C_c^\infty(\mathbb{R}_+^{n+1};\mathbb{R}_0^n,\mu)$ will be denoted by $F = (f;g_1,\ldots,g_\mu)$. Similarly defined are the spaces $C^\infty(\mathbb{R}_+^{n+1};\mathbb{R}_0^n,\mu)$, $C_c^\infty(\Omega\cup\omega;\omega,\mu)$ and $C^\infty(\Omega\cup\omega;\omega,\mu)$. For $(P;Q_1,\ldots,Q_\mu)$ given partial differential operators we define

$$P:u \in C^\infty(\Omega\cup\omega) \to (Pu;Q_1u,\ldots,Q_\mu u) \in C^\infty(\Omega\cup\omega;\omega,\mu)$$

where $Q_\nu u$ denotes $Q_\nu(D,D_t)u\big|_\omega$, $1 \le \nu \le \mu$. We define an operator $E$ from $C_c^\infty(\Omega\cup\omega;\omega,\mu)$ into $C^\infty(\Omega\cup\omega)$ as follows:

$$EF = K_*f + \sum_{\nu=1}^{\mu} K_{\nu*}'g_\nu$$

where $*$ indicates the convolution in $\mathbb{R}^{n+1}$ and $*'$ the convolution in $\mathbb{R}^n$. From properties (2.60) and (2.61) of the kernels it follows

$$PEF = F-LF$$

where $LF = (\beta(x)*'f(x,0);\beta(x)*'g_1(x),\ldots,\beta(x)*'g_\mu(x))$. We call $E$ a *parametrix* of the regular boundary-value problem defined by $(P;Q_1,\ldots,Q_\mu)$ and $L$ a smoothing operator.

Theorem 2.16. *Suppose that the distributions* $K(x,t)$ *and* $(K_\nu(x,t))_{1\le\nu\le\mu}$ *defined by* (2.58) *and* (2.59) *and satisfying* (2.60) *and* (2.61) *belong to* $\Gamma^{(d_1,\ldots,d_{n+1})}(\overline{\mathbb{R}_+^{n+1}}\setminus\{0\})$. *Then* $(P;Q_1,\ldots,Q_\mu)$ *defines a regular* $(d_1,\ldots,d_{n+1})$-*hypoelliptic boundary problem.*

*Proof.* Let $u \in C^k(\Omega\cup\omega)$ be a solution of (2.5) with $f \in \Gamma^{(d_1,\ldots,d_{n+1})}(\Omega\cup\omega)$ and $g_\nu \in \Gamma^{(d_1,\ldots,d_n)}(\omega)$, $1 \le \nu \le \mu$. Let $\Omega'$ be a bounded open set contained with a plane piece of boundary $\omega' \subset \omega$ and such that $\overline{\Omega'\cup\omega'} \subset \overline{\Omega\cup\omega}$. Let $\alpha \in C_c^\infty(\Omega\cup\omega)$ be such that $\alpha = 1$ on $\overline{\Omega'\cup\omega'}$. We have

$$P(\alpha u) = (g;h_1,\ldots,h_\mu) \tag{2.90}$$

with $g \in C_c(\Omega\cup\omega)$ belonging to $\Gamma^{(d_1,\ldots,d_{n+1})}(\Omega'\cup\omega')$ and $h_\nu \in C_c(\omega)$ belonging $\Gamma^{(d_1,\ldots,d_n)}(\omega')$, $1 \le \nu \le \mu$. Since, by assumption, $K$ and $(K_\nu)$ belong to $\Gamma^{(d_1,\ldots,d_{n+1})}(\mathbb{R}_+^{n+1}\setminus\{0\})$, a standard argument about convolution of distribution shows that $EP(\alpha u) \in \Gamma^d(\Omega'\cup\omega')$. In view of the definitions of $E$, $K$ and $K_\nu$, we have:

$$EP(\alpha u)^\wedge = (1-\chi(\xi))G(\xi,t)\hat{g}(\xi,t) + \sum_{\nu=1}^{\mu}(1-\chi(\xi))H_\nu(\xi,t)\hat{h}_\nu(\xi)$$

From (2.90) it follows that $(\alpha u)\hat{}$ is the solution of the initial-value problem

$$\begin{cases} P(\xi,D_t)(\alpha u)\hat{}(\xi,t) = \hat{g}(\xi,t) \\ Q_\nu(\xi,D_t)(\alpha u)\hat{}(\xi,t)\Big|_{t=0} = \hat{h}(\xi), \quad 1 \le \nu \le \mu . \end{cases}$$

By the representation formula (2.31) and the definition of $G(\xi,t)$ and $H_\nu(\xi,t)$, we get

$$(1-\chi(\xi))(\alpha u)\hat{}(\xi,t) = EP(\alpha u)\hat{}.$$

Thus,

$$\alpha u = EP(\alpha u) - \beta *'(\alpha u)$$

where $\beta(x)$, the inverse Fourier transform of $\chi(\xi)$, is an entire analytic function. As we have remarked above, $EP(\alpha u) \in \Gamma^d(\Omega' \cup \omega')$ and, since $\beta(x)$ is analytic, the convolution $\beta *'(\alpha u)(x,t)$ is analytic in x and of class $C^k$ with respect to t. Finally, by using the equation $P(D,D_t)u = f$ and the assumption $f \in \Gamma^{(d_1,\ldots,d_{n+1})}(\Omega \cup \omega)$ we easily conclude that $u \in \Gamma^{(d_1,\ldots,d_{n+1})}(\Omega' \cup \omega')$. Since the choice of $\Omega'$ and $\omega'$ is arbitrary, the theorem follows. Q.E.D.

The next theorem summarizes the results of this chapter:

Theorem 2.17. *The following are equivalent conditions:*

*(1)* $(P;Q_1,\ldots,Q_\mu)$ *defines a regular* $(d_1,\ldots,d_{n+1})$*-hypoelliptic boundary problem.*

*(2) For some open set* $\Omega \subset \mathbb{R}^{n+1}_+$ *with a plane piece of boundary* $\omega \subset \mathbb{R}^n_0$, *all solutions* $u \in C^k(\Omega \cup \omega)$ *of the problem*

$$\begin{cases} P(D,D_t)u = 0 \quad \text{in} \quad \Omega \cup \omega \\ Q_\nu(D,D_t)u\Big|_\omega = 0, \quad 1 \le \nu \le \mu, \end{cases}$$

*belong to* $\Gamma^{(d_1,\ldots,d_{n+1})}(\Omega \cup \omega)$;

*(3) There is a constant* $C > 0$ *such that the set*

$$D = \{\zeta \in C^n : [\operatorname{Re} \zeta]_{d'} \ge C(1+|\operatorname{Im} \zeta|)\}$$

*is contained in* A *and the characteristic function* $C(\zeta)$ *does not vanish on* D;

*(4) There are distributions* $K(x,t)$ *and* $(K_\nu(x,t))_{1\le\nu\le\mu}$ *belonging to*

$\Gamma^{(d_1,\ldots,d_{n+1})}(\overline{\mathbb{R}^{n+1}_+} \setminus \{0\})$ *and satisfying the boundary problems* (2.60) *and* (2.61).

*(5) There is a parametrix E mapping* $C_c^\infty(\Omega \cup \omega;\omega,\mu)$ *into* $C^\infty(\Omega \cup \omega)$ *such that, for all* $F \in C_c^\infty(\Omega \cup \omega;\omega,\mu)$,

$$PEF = F - LF$$

*with L a smoothing operator. Moreover, if* $F = (f;g_1,\ldots,g_\mu)$ *with* $f \in \Gamma^{(d_1,\ldots,d_{n+1})}(\Omega \cup \omega)$ *and* $g_\nu \in \Gamma^{(d_1,\ldots,d_n)}(\omega)$, $1 \leq \nu \leq \mu$, *then* $EF \in \Gamma^{(d_1,\ldots,d_{n+1})}(\Omega \cup \omega)$.

# CHAPTER 3
# REGULAR SEMIELLIPTIC BOUNDARY-VALUE PROBLEMS

## 3.1 INTRODUCTION

Let $P(D,D_t)$ be a semielliptic operator in the form

$$P(D,D_t) = D_t^{\sigma} + \sum_{j=1}^{\sigma} \alpha_j(D) D_t^{\sigma-j}$$

where, in general, $\sigma \leq$ order P. We will also assume $P^0(D,D_t)$, the principal part of P [Section 1.5] is in the form

$$P^0(D,D_t) = D_t^{\sigma} + \sum_{j=1}^{\sigma} \alpha_j(D) D_t^{\sigma-j}$$

As stated in Chapter 1, $P(D,D_t)$ is hypoelliptic with index of hypoellipticity $(d_1,\ldots,d_{n+1})$ where:

$$d_j = \frac{m}{m_j}, \quad 1 \leq j \leq n+1,$$

$$m_j = \text{degree of } P(D) \text{ as a polynomial in } D_j, \ 1 \leq j \leq n,$$

$$m_{n+1} = \sigma$$

$$m = \max_{1 \leq j \leq n+1} m_j .$$

Finally, we assume $P(D,D_t)$ is of type $\mu$ with $\mu \leq \sigma$. Recall that this means that there is a compact set $K \subset \mathbb{R}^n$ such that, for all $\xi \in \mathbb{R}^n \setminus K$, $P(\xi,t) = 0$, as a polynomial in $\tau$, has $\mu$ roots with positive imaginary part and none that are real.

Suppose that $\mu$ partial differential operators with constant coefficients $Q_1(D,D_t),\ldots,Q_\mu(D,D_t)$ are given in the form

$$
\begin{aligned}
Q_\nu(D,D_t) &= Q_\nu^0(D,D_t) + \ldots \\
Q_\nu^0(D,D_t) &= \sum_{\langle p,d\rangle = n_\nu} a_{p,\nu} D^{p'} D_t^{p_{n+1}}, \quad 1 \le \nu \le \mu.
\end{aligned}
\tag{3.1}
$$

with $n_\nu$ nonnegative integers. Let $\Omega$ be an open set in $\mathbb{R}_+^{n+1}$ with plane piece of boundary $\omega$ as described in Chapter 2. In this chapter the boundary-value problems $(P;Q_1,\ldots,Q_\mu)$ and $(P^0;Q_1^0,\ldots,Q_\mu^0)$ are studied.

First the case $P = P^0$, $Q_j = Q_j^0$ will be considered under the assumption that each $d_j$ is an integer (necessarily $\ge 1$) for $1 \le j \le n+1$. In this case specific information about the fundamental solutions to $(P^0;Q_\nu^0)_{1\le\nu\le\mu}$ which generalize the well known results for elliptic boundary-value problems are obtained (see [2]).

There are large classes of operators for which the $d_j$ are all integers. For example let $P(D,D_t)$ be a semielliptic operator satisfying:

$(kp)_1$ $\quad m_1 = m_2 = m_3 = \ldots = m_n = km_{n+1}$,

$(kp)_2$ $\quad$ there is a $\delta > 0$ such that for all $(\xi_1,\ldots,\xi_n)$ such that $\xi_1^2+\ldots+\xi_n^2 = 1$, $\tau \in \mathbb{C}$, $P^0(\xi_1,\ldots,\xi_n,\tau) = 0 \Rightarrow \operatorname{Im}\tau \ge \delta$. In this case

$$
d_i = m/m_i = km_{n+1}/m_i = \begin{cases} 1 & 1 \le i \le n \\ k & i = n+1 \end{cases}
$$

(Note: It is an easy exercise to show $(kp)_1$ and $(kp)_2$ imply k is even.) Thus $P(D,D_t)$ is $(1,1,\ldots,k)$-hypoelliptic. Any semielliptic operator satisfying $(kp)_1$ and $(kp)_2$ is called k-*parabolic*.

Example. The heat operator $\frac{\partial}{\partial t} - (\frac{\partial}{\partial x_1})^2 + \ldots + (\frac{\partial}{\partial x_n})^2$ is 2-parabolic.

In the general case ($d_j$ not necessarily integers) we will show that the regularity of the solutions to $(P,Q_1,\ldots,Q_\mu)$ can be characterized in terms of the characteristic function, $C^0(\xi)$, of the associated problem $(P^0;Q_1^0,\ldots,Q_\mu^0)$. In fact, we will prove the following

Theorem 3.1. *(a) If* $C^0(\xi) \ne 0$ *for all* $\xi \in \mathbb{R}^n \setminus \{0\}$ *then the boundary-value problem* $(P;Q_1,\ldots,Q_\mu)$ *in* $\Omega \cup \omega$, *is regular semielliptic**.

---

*$(P;Q_1,\ldots,Q_\mu)$ is *regular semielliptic* if P is semielliptic and $(P;Q_1,\ldots,Q_\mu)$ is regular hypoelliptic.

*(b) Conversely, if* $(P;Q_1,\ldots,Q_\mu)$ *defines a regular semielliptic boundary-value problem in* $\Omega \cup \omega$ *then either* $C^0(\xi) = 0$ *identically or* $C^0(\xi) \neq 0$ *for* $\xi \in \mathbb{R}^n \setminus \{0\}$.

When $P(D,D_t)$ is elliptic this gives the result obtained for elliptic boundary-value problems in Hormander [11], Theorem 3.3, page 241.

## 3.2 SEMIHOMOGENEOUS OPERATORS

First let $P = P^0$, $Q_1 = Q_1^0,\ldots,Q = Q^0$ then the following theorem holds whether the $d_j$'s are integers or not.

Theorem 3.2. *The following are equivalent:*

*(1)* $(P;Q_1,\ldots,Q_\mu)$ *defines a regular semielliptic boundary-value problem in* $\Omega \cup \omega$

*(2)* $C(\xi) \neq 0$ *for all* $\xi \in \mathbb{R}^n \setminus \{0\}$.

*Proof.* (1 => 2). Since $(P;Q_1,\ldots,Q_\mu)$ is semielliptic all solutions in $C^k(\Omega \cup \omega)$ of

$$\begin{cases} P(D,D_t)u = 0 \text{ in } \Omega \cup \omega \\ Q_\nu(D,D_t)u = 0 \text{ in } \omega \end{cases}$$

belonging to $\Gamma^d(\Omega \cup \omega)$. One can proceed as in Theorem 2.12 to show that as a consequence of this there exists a constant $M > 0$ such that if

$$[\zeta]_{d'} \geq M(1+|\operatorname{Im} \zeta|) \tag{3.2}$$

then $\zeta \in A$ and $C(\zeta) \neq 0$, where $A$ is the same as in Section 2.2. For $\zeta \in A$ let $\tau_1(\zeta),\ldots,\tau_\mu(\zeta)$ be the $\mu$-roots of $P(\zeta,\tau) = 0$ with positive imaginary parts. Since $P(\zeta,\tau)$ is semihomogeneous of degree $\bar{m}$ with respect to $(d_1,\ldots,d_{n+1})$ it follows that each root $\tau_j(\zeta)$ is semihomogeneous of degree $d_{n+1}$ with respect to $(d_1,\ldots,d_n)$. Using this and the fact that each boundary operator is semihomogeneous of degree $n_\nu$ for each $\nu = 1,\ldots,\mu$, an easy calculation shows that the characteristic function of $(P;Q_1,\ldots,Q_\mu)$ defined for each $\zeta \in A$ by (2.15) is semihomogeneous of degree

$$n_1 + n_2+\ldots+n_\mu-d_{n+1}(1+2+\ldots+(\mu-1))$$

with respect to $(d_1,\ldots,d_n)$.

It follows from (3.2) that on the surface

$$[\xi]_{d'} = M, \qquad \xi \in \mathbb{R}^n$$

$C(\xi) \neq 0$. Now for any $\xi \in \mathbb{R}^n \setminus \{0\}$ the vector

$$z = \left( \frac{\xi_1}{\left(\frac{[\xi]_d}{M}\right)^{d_1}}, \ldots, \frac{\xi_n}{\left(\frac{[\xi]_d}{M}\right)^{d_n}} \right)$$

satisfies $[z]_{d'} = M$ and hence $C(z) \neq 0$. The semihomogeneity of $C(\xi)$ implies that $C(\xi) \neq 0$ for all $\xi \in \mathbb{R}^n \setminus \{0\}$.

(2 => 1). Assuming $C(\xi) \neq 0$ for $\xi \in \mathbb{R}^n \setminus \{0\}$, we form the functions, $H_\nu(\xi,t)$, $G(\xi,t)$ defined by (2.54) and (2.55), respectively. For $\xi \neq 0$ they satisfy

$$\begin{cases} P(\xi,D_t)G(\xi,t) = \delta(t) \\ Q_\nu(\xi,D_t)G(\xi,0) = 0 \end{cases}$$

$$\begin{cases} P(\xi,D_t)H_j(\xi,t) = 0 & 1 \leq \nu \leq \mu, \\ Q_\nu(\xi,D_t)H_j(\xi,0) = \delta_{j,\nu}, & 1 \leq j \leq \mu. \end{cases} \tag{3.3}$$

As in Section 2.8 we can show that in this case for all $\xi \neq 0$ $G(\xi,t)$ and $H_\nu(\xi,t)$ define tempered distributions in $\mathbb{R}^n \setminus \{0\}$ depending on t. We thus define

$$\begin{aligned} K_0(x,t) &= F_\xi^{-1}G(\xi,t) \\ K_\nu(x,t) &= F_\xi^{-1}H_\nu(\xi,t), \quad t \geq 0, \quad \xi \neq 0. \end{aligned} \tag{3.4}$$

By taking the inverse Fourier transform of (3.4) we see $K_0(x,t)$ and $(K_\nu(x,t))_{1 \leq \nu \leq \mu}$ satisfy

$$\begin{cases} P(D,D_t)K_0(x,t) = \delta(x,t) \text{ in } \mathbb{R}_+^{n+1} \\ Q_\nu(D,D_t)K_0(x,0) = 0 \text{ in } \mathbb{R}^n \end{cases} \tag{3.5}$$

and

$$\begin{cases} P(D,D_t)K_j(x,t) = 0 \text{ in } \mathbb{R}_+^{n+1} \\ Q_\nu(D,D_t)K_j(x,0) = \delta_{\nu,j}\delta(x) \text{ in } \mathbb{R}^n \end{cases} \tag{3.6}$$

$$1 \leq \nu \leq \mu, \quad 1 \leq j \leq \mu.$$

Thus in this case $K_0(x,t)$, $K_1(x,t),\ldots,K_\mu(x,t)$ form a fundamental solution of the semielliptic boundary-value problem $(P;Q_1,\ldots,Q_\mu)$. One can now proceed to find estimates on the derivatives of $K_j(x,t)$ in compact subsets of $\Omega \cup \omega$ to show that $K_j(x,t) \in \Gamma^d(\Omega \cup \omega)$ as in Chapter 2 and by taking proper convolutions show that $(P;Q_1,\ldots,Q_\mu)$ is regular semielliptic so that the theorem follows. However, in the case when $(d_1,\ldots,d_{n+1})$ are integers, the $K_j$'s define semihomogeneous distributions as will be shown in Section 3.4 and using the properties of such distributions the regularity of the $K_j$'s will be established.

## 3.3 SEMIHOMOGENEOUS DISTRIBUTIONS

Definition 3.1. *Let* $T \in \mathcal{D}'(\mathbb{R}^n)$ *then* $T$ *is semihomogeneous of degree* $\alpha \in C$ *with respect to* $(d_1,\ldots,d_n)$ *if for each* $\lambda > 0$

$$\langle T(x), \phi(\lambda^{-d_1}x_1,\ldots,\lambda^{-d_n}x_n)\rangle$$
$$= \lambda^{|d|+\alpha}\langle T(x),\phi(x)\rangle$$

*holds for every* $\phi(x) \in C_c^\infty(\mathbb{R}^n)$.

It is a simple exercise to show that when $T(x)$ is a locally integrable function then this definition coincides with definition 1.9.

Let $\sum_n = \{x \in \mathbb{R}^n : \rho(x) = 1\}$, where $\rho(x)$ is the metric introduced in Section 2.10. If $T(x)$ is a $C^\infty$ function on $\sum_n$ then for each $\alpha \in C$ we can extend $T(x)$ to a function $T_\alpha(x)$ on $\mathbb{R}^n \setminus \{0\}$ which coincides with $T$ on $\sum_n$ and is semihomogeneous of degree $\alpha$ with respect to $(d_1,\ldots,d_n)$. In fact, since

$$\rho(\lambda^{d_1}x_1,\ldots,\lambda^{d_n}x_n) = \lambda\rho(x_1,\ldots,x_n)$$

then one can let

$$T_\alpha(x) = \rho(x)T\left(\frac{x_1}{\rho(x)^{d_1}},\ldots,\frac{x_n}{\rho(x)^{d_n}}\right), \quad \forall x \text{ in } \mathbb{R}^n \setminus \{0\}. \tag{3.7}$$

When $d_1,\ldots,d_n$ are integers $\geq 1$ one can show that $T_\alpha(x)$ defines a tempered distribution $\mathbb{R}^n$. To see this let $\phi(x) \in \mathcal{S}(\mathbb{R}^n)$ and formally write the integral

$$\int_{\mathbb{R}^n} T_\alpha(x)\phi(x)\,dx = \int_0^\infty \int_{\sum_n} T_\alpha(\rho^d\omega)\phi(\rho^d\omega)\rho^{|d|-1}\,d\rho\, d\omega$$

Since $T_\alpha(\rho^d\omega) = \rho^\alpha T(\omega)$,

$$\begin{aligned}\int_{\mathbb{R}^n} T_\alpha(x)\phi(x)\,dx &= \int_0^\infty \int_{\sum_n} T(\omega)\phi(\rho^d\omega)\rho^{\alpha+|d|-1}\,d\rho\, d\omega \\ &= \int_0^\infty h_\phi(\rho)\rho^{-\alpha'-1}\,d\rho\end{aligned} \tag{3.8}$$

where $\alpha' = -\alpha-|d|$, $\omega \in \sum_n$, $d\omega$ is the surface element on $\sum_n$ and

$$h_\phi(\rho) = \int_{\sum_n} T(\omega)\phi(\rho^d\omega)\,d\omega.$$

Notice that since the $d_j$'s are integers $\geq 1$ $h_\phi(\rho)$ belongs to $S'(\mathbb{R}^+)$ and so the last integral in (3.8) behaves at infinity. To give it a meaning at the origin, consider two cases. First, if $\alpha'$ is not a positive integer then integrating by parts p + 1 times in 3.8 gives

$$<T_\alpha(x),\phi> = [\alpha'(\alpha'-1)\ldots(\alpha'-p)]^{-1}\int_0^\infty \rho^{-\alpha'+p} h_\phi^{(p+1)}(\rho)\,d\rho \tag{3.9}$$

For p sufficiently large the last integral converges, defining $T_\alpha(x)$ as an element of $S'(\mathbb{R}^n)$. Denote this distribution by pf.$T_\alpha$. It is obvious that pf.$T_\alpha = T_\alpha$ on $\mathbb{R}^n \setminus \{0\}$.

The association $\alpha \to$ pf.$T_\alpha$ taking $C \to S'(\mathbb{R}^n)$ defines a meromorphic function with poles $\{\alpha \in C{:}\alpha' \in \mathbb{N}\}$. If $\alpha'$ is equal to p we define a tempered distribution by using the finite part of the integral (3.8). This can most easily be done by using the constant term in the Laurent expansion of $<$pf. $T_\alpha,\phi>$ treated as a function of a complex variable $\alpha$, when $\alpha'$ is in a neighborhood of p. Consider the expression in (3.9)

$$A(\alpha') = \rho^{-\alpha'+p}[\alpha'(\alpha'-1)\ldots(\alpha'-p)]^{-1}.$$

The constant term in the expansion of $A(\alpha')$ is the coefficient of $(\alpha'-p)$ in the expansion of $(\alpha'-p)A(\alpha')$ i.e.

$$(\alpha'-p)A(\alpha') = \rho^{-\alpha'+p}[\alpha'(\alpha'-1)\ldots(\alpha'-p+1)]^{-1}.$$

The coefficient of $(\alpha'-p)$ in the Laurent expansion of $(\alpha'-p)A(\alpha')$ is just

$$\frac{d}{d\alpha'}[(\alpha'-p)A(\alpha')]\Big|_{\alpha' = p}$$

A simple computation shows that this derivative is

$$\frac{1}{p!}(\log\frac{1}{\rho} - 1 - \frac{1}{2}\ldots-\frac{1}{p}).$$

Thus for $\alpha' \in \mathbb{N}$ take for the definition of pf.$T_\alpha$

$$\begin{aligned} <\text{pf.}T_\alpha,\phi> &= \text{pf.}\int_0^\infty h_\phi(\rho)\rho^{-p-1}d\rho \\ &= \int_0^\infty \frac{1}{p!}(\log\frac{1}{\rho} - 1 - \frac{1}{2}\ldots\frac{1}{p})h_\phi^{(p+1)}(\rho)\,d\rho. \end{aligned} \tag{3.10}$$

To sum up we take the definition of $\text{pf}.T_\alpha(x)$ to be (3.8) if $\alpha' \in \mathbb{N}$ and (3.10) when $\alpha' \in \mathbb{N}$.

Proposition 3.1. *If* $T$ *is a* $C^\infty$ *function in* $\mathbb{R}^n \setminus \{0\}$ *semihomogeneous of degree* $\alpha$ *with respect to* $(d_1,\dots,d_n)$ *and* $\alpha' \notin \mathbb{N}$, *then* $\text{pf}.T_\alpha$ *is the only distribution on* $\mathbb{R}^n$ *semihomogeneous of the same degree which coincides with* $T_\alpha$ *in* $\mathbb{R}^n \setminus \{0\}$.

*Proof.* Everything is obvious from what was said above. The semihomogeneity follows from the calculation

$$<\text{pf}.T_\alpha(x), \phi(\lambda^{-d}\chi)> = \int_0^\infty \int_{\Sigma_n} T(\omega)\,\phi\left(\left(\frac{\rho}{\omega}\right)^d \omega\right)\rho^{\alpha+|d|-1}\,d\rho\, d\omega.$$

Make the change of variable $r = \frac{\rho}{\lambda}$ and one obtains

$$\int_0^\infty \int_\Sigma T(\omega)\,\phi(r^d\omega)\,\lambda^{\alpha+|d|-1} r^{\alpha+|d|-1}\lambda\, d\rho\, d\omega$$

$$= \lambda^{\alpha+|d|} <\text{pf}.T_\alpha(x), \phi(x)>$$

Q.E.D.

In general when $\alpha' \in \mathbb{N}$ $\text{pf}.T_\alpha$ is not semihomogeneous. Let $\alpha' \in \mathbb{N}$, then

$$<\text{pf}.T_\alpha, \phi(\lambda^{-d}\chi)> = \int_{\Sigma_n} \int_0^\infty T(\omega)\left(\log \frac{1}{\rho} - 1 - \frac{1}{2} - \dots - \frac{1}{p}\right)\frac{d^{p+1}}{d\rho^{p+1}}\phi(\rho\lambda^{-d}\omega)\, d\rho\, d\omega.$$

Making the change of variable $\rho \to \lambda\rho$, the above integral equals

$$\lambda^p \Big[\int_{\Sigma_n} \int_0^\infty T(\omega)\log \frac{1}{\rho} - 1 - \frac{1}{2} - \dots - \frac{1}{p}\Big)\frac{d^{p+1}}{d\rho^{p+1}}\phi(\rho^d\omega)\, d\rho\, d\omega$$

$$+ \lambda^p \log \lambda \int_{\Sigma_n} T(\omega) \int_0^\infty \frac{d^{p+1}}{d\rho^{p+1}}\phi(\rho^d\omega)\, d\rho\, d\omega.$$

The second term in the above integral is equal to

$$\lambda^p \log \lambda \int_{\Sigma_n} T(\omega)\frac{d^p}{d\rho^p}\phi(\rho^d\omega)\Big|_{\rho=0}\, d\omega.$$

We will show that

$$\frac{d^p}{d\rho^p}\phi(\rho^d\omega)\Big|_{\rho=0} = \sum_{<\nu, d>=p} c_\nu w^\nu$$

with $c_\nu$ constants. Hence it follows that

*If* $\alpha' \in \mathbb{N}$, *then* pf.$T_\alpha$ *is semihomogeneous of degree* $\alpha$ *with respect to* $(d_1,\ldots,d_n)$ *if and only if*

$$\int_{\Sigma_n} T(\omega)\omega^\nu d\omega = 0$$

*for all* $\nu$ *such that* $\langle\nu,d\rangle = p$.

Therefore, consider

$$\frac{d^p}{d\rho^p}(\phi(\rho^d\omega))\Big|_{\rho=0}.$$

That this is of the desired form follows from

Lemma 3.1. *If* F *is a* $C^\infty$ *function then*

$$\frac{d}{d\rho^p}(F(\rho^d\omega))\Big|_{\rho=0}$$

*is a semihomogeneous polynomial in* $\omega$ *of degree* p *with respect to* $(d_1,\ldots,d_n)$.

*Proof.* We prove this by induction on the order of the differentiation. For p = 1

$$\frac{d}{d}(F(\rho^d\omega)) = \sum_{j=1}^{n} \frac{\partial F}{\partial x_j}\, d_j\, \rho^{d_j-1}\, \omega_j.$$

The only terms in this sum which do not vanish at $\rho = 0$ are those whose index j are such that $d_j = 1$. Therefore, $\frac{d}{d\rho}F(\rho^d\omega)\Big|_{\rho=0}$ is semihomogeneous of degree 1.

Now suppose the lemma is true for all orders of differentiation less than p. Write

$$\frac{d^p}{d\rho^p}F(\rho^d\omega)\Big|_{\rho=0} = \frac{d^{p-1}}{d\rho^{p-1}}\Big(\sum_{j=1}^{n}\frac{\partial F}{\partial x_j}\, d_j\, \rho^{d_j-1}\, \omega_j\Big)\Big|_{\rho=0}$$

$$= \sum_{s+t=p-1} \frac{1}{s!t!}\, \frac{d^s}{d_\rho{}^s}\Big(\frac{\partial F}{\partial x_j}\Big)\Big|_{\rho=0}\, \omega_j d_j\, \frac{d^t}{d_\rho{}^t}\, \rho^{d_j-1}\Big|_{\rho=0}$$

Using the induction hypothesis on

$$\frac{d^s}{d\rho^s}\Big(\frac{\partial F}{\partial x_j}\Big)\Big|_{\rho=0}, \quad s < p$$

with F replaced by $(\frac{\partial F}{\partial x_j})$, $\frac{d^s}{d\rho^s}(\frac{\partial F}{\partial x_j})_{\rho=0}$ is semihomogeneous of degree s.

It suffices to show that $d_j\omega_j \frac{d^t}{d\rho^t}\rho^{d_j-1}\Big|_{\rho=0}$ is semihomogeneous of degree $t + 1$. Now

$$d_j\omega_j \frac{d^t}{d\rho^t}\rho^{d_j-1} = d_j(d_j-1)\ldots(d_j-t)\rho^{d_h-(t+1)}\omega_j \;.$$

It is easy to see that the only indices for which this expression is non zero at $\rho = 0$ are those j such that $d_j = t + 1$. Since this is a polynomial of degree 1 it must be by semihomogeneous of degree $t + 1$.

Since pf. $T_\alpha$ is in $S'(\mathbb{R}^n)$ we define the Fourier transform $(\text{pf}.T_\alpha)^\wedge$ of $\text{pf}.T_\alpha$ by

$$<(\text{pf}.T_\alpha)^\wedge, \phi(x)> = <\text{pf}.T_\alpha, \hat{\phi}(\xi)> \;.$$

For $\alpha' \varepsilon \mathbb{N}$

$$<(\text{pf}.T_\alpha)^\wedge, \phi> = [\alpha'(\alpha'-1)\ldots(\alpha'-p)]^{-1}\int_0^\infty \rho^{-\alpha'+p} h_{\hat{\phi}}^{(p+1)}(\rho)\,d\rho \tag{3.11}$$

and for $\alpha' \varepsilon \mathbb{N}$

$$<(\text{pf}.T_\alpha)^\wedge, \phi> = \int_0^\infty \frac{1}{p!}(\log\frac{1}{\rho} - 1 - \frac{1}{2} - \ldots - \frac{1}{p}) h_{\hat{\phi}}^{(p+1)}(\rho)\,d\rho \;. \tag{3.12}$$

Proposition 3.2. *If* $T(x)$ *is a semihomogeneous distribution in* $S'(\mathbb{R}^n)$ *of degree* $\alpha$ *with respect to* $(d_1,\ldots,d_n)$ *then* $\hat{T}(\xi)$ *is semihomogeneous of degree* $\alpha'$ *with respect to* $(d_1,\ldots,d_n)$.

*Proof.* $<\hat{T}_\alpha(x), \phi(\lambda^{-d_1}x_1,\ldots,\lambda^{-d_n}x_n)>$

$$= <T_\alpha(\xi), \hat{\phi}(\lambda^{-d_1}x_1,\ldots,\lambda^{-d_n}x_n)(\xi)>$$

$$= <T_\alpha(\xi), \int_{\mathbb{R}^n}\phi(\lambda^{-d_1}x_1,\ldots,\lambda^{-d_n}x_n)e^{-i<x,\xi>}dx>$$

$$= <T_\alpha(\xi), \lambda^{|d|}\int_{\mathbb{R}^n}\phi(x_1,\ldots,x_n)e^{-i<x,(\lambda^{d_1}\xi_1,\ldots,\lambda^{d_n}\xi_n)>}dx$$

$$= \lambda^{|d|}<T_\alpha(\xi), \hat{\phi}(\lambda^{d_1}\xi_1,\ldots,\lambda^{d_n}\xi_n)>$$

$$= \lambda^{|d|}\lambda^{-\alpha-|d|}<\hat{T}_\alpha(x), \phi(x)> = \lambda^{\alpha'+|d|}<\hat{T}_\alpha(x), \phi(x)>.$$

Proposition 3.3. *Let* T *be the function defined above semihomogeneous of degree* $\alpha$ *with respect to* $(d_1,\ldots,d_n)$ *and* $C^\infty$ *in* $\mathbb{R}^n \setminus \{0\}$. *Then* $\hat{T}(x)$ *belongs to* $\Gamma^d(\mathbb{R}^n \setminus \{0\})$.

*Proof.* Let $\psi(\xi)$ be in $C_c^\infty(\mathbb{R}^n)$ such that $\psi = 1$ for $|\xi| < 1$ and $\psi = 0$ for $|\xi^2 > 2$. Write $T = (\psi T)\hat{} + ((1-\psi)T)\hat{}$. Now since $\psi T$ has compact support, $(\psi T)\hat{}$ is analytic in $\mathbb{R}^n$ and hence in $\Gamma^d(\mathbb{R}^n)$. So one must show $((1-\psi)T)\hat{} \in \Gamma^d(\mathbb{R}^n \setminus \{0\})$. For any n-tuple of nonnegative integers p and q consider

$$x^q D^p((1-\psi)T)\hat{}(x).$$

Using the Leibniz rule for derivatives:

$$|x^q D^p((1-\psi)T)\hat{}(x)| \le |T_0| + \sum_{\substack{|r|+|s|+|t|=|q| \\ s\neq 0}} |T_{r,s,t}|$$

where

$$T_0 = \sum_{|\gamma|+|\beta|=|q|} \frac{q!}{(2\pi)^n \gamma!\beta!} \int_{\mathbb{R}^n} [1-\psi(\xi)] D^\gamma \xi^p D^\beta T(\xi)\, d\xi, \tag{3.13}$$

and

$$T_{r,s,t} = \frac{q!}{(2\pi)^n r!s!t!} \int_{\mathbb{R}^n} D^r \xi^p D^s (1-\psi(\xi)) D^t(\xi)\, dt. \tag{3.14}$$

Since $|\xi_i| \le \rho(\xi)^{d_i}$ and $|D^\gamma \xi^p| \le 2^{|p|} |\xi^{p-\gamma}$, it follows that there exists a constant $c > 0$ such that

$$|T_0| \le c^{|p|+1} \int_{\Sigma_n} \int^\infty \rho^{<d,p>-<\gamma,d>-<\beta,d>+\alpha+|d|-1} |D^\beta T(y)|\, d\rho\, d\sigma(y). \tag{3.15}$$

To estimate $T_{r,s,t}$, we see that since $s \neq 0$, the support of $D^s(1-\psi)$, is contained in a compact set H. Letting

$$B = \sup_{\xi \varepsilon H} |D^r \xi^p|$$

we have

$$\begin{aligned} |T_{r,s,t}| &\le q!\ B \int_H D^s(1-\psi) D^t T(\xi)\, d\xi \\ &\le q!\ C \end{aligned} \tag{3.16}$$

for some $C > 0$.

The integral in (3.15) converges provided

$$\sum d_j p_j - |q| + k + \sum d_j - 1 < -1.$$

Since $\langle q,d\rangle > |q|$, this condition is satisfied if

$$|q| > \sum d_j p_j + \sum d_j + \alpha. \tag{3.17}$$

Assuming this, (3.15) and (3.16) imply

$$|x^q D^p((1-\psi)T)\hat{}(x)| \le C_1^{|p|+1}|q|! \tag{3.18}$$

for some $C_1 > 0$. If we further assume

$$|q| < \sum_{j=1}^{n} p_j d_j + \sum_{p=1}^{n} d_j + \alpha + 1 \tag{3.19}$$

then $|q|! \le \Gamma(\sum_{j=1}^{n} p_j d_j + \sum_{j=1}^{n} d_j + \alpha + 1)$

$$\le C_2^{|p|+1}(p_1!)^{d_1}\cdots(p_2!)^{d_n}. \tag{3.20}$$

Therefore (3.18) and (3.20) imply that if K is a compact subset of $\mathbb{R}^n \setminus \{0\}$, there exists a constant $C > 0$ such that

$$\sup_{x\in K} |D^p((1-\psi)T)\hat{}(x)| \le C^{|p|+1}(p_1!)^{d_1}\cdots(p_n!)^{d_n}.$$

It follows from Proposition 3.1 and 3.2 that if $\alpha' \in \mathbb{N}$ then $(\text{pf}.T_\alpha)\hat{}(x)$ is a semihomogeneous distribution which is in $\Gamma^d(\mathbb{R}^n \setminus \{0\})$. In fact we shall see that $(\text{pf}.T_\alpha)\hat{}(x)$ coincides with a function. Indeed, $(\text{pf}.T_\alpha)\hat{}(x)$ must be the sum of a $C^\infty$ function in $\mathbb{R}^n \setminus \{0\}$ and a semihomogeneous distribution $S(x)$ supported at the origin in $\mathbb{R}^n$. But then

$$S(x) = \sum_{\langle \ell,d\rangle=\alpha'} C^\ell \delta^{(\ell)}(x).$$

Now $\langle S(x),\phi(\lambda^{-d}x)\rangle = \sum_{\langle \ell,d\rangle=\alpha'} C^\ell \lambda^{-\langle \ell,d\rangle} D^\ell \phi(0)$

$$= \sum_{\langle \ell,d\rangle=\alpha'} C^\ell \lambda^{-\alpha'} D^\ell \phi(0).$$

On the other hand,

$$<S(x),\phi(\lambda^{-d_1}x_1,\ldots,\lambda^{-d_n}x_n)>$$

$$= \lambda^{\alpha'+|d|} \sum_{<\ell,d>=\alpha'} C^\ell D^\ell \phi(0).$$

Therefore,

$$\lambda^{\alpha'+|d|} \sum_{<\ell,d>=\alpha'} C^\ell D^\ell \phi(0) = \sum_{<\ell,d>=\alpha'} C^\ell \lambda^{-\alpha'} D \quad (0)$$

or

$$\lambda^{2\alpha'+|d|} \sum_{<\ell,d>=\alpha'} C^\ell D^\ell \phi(0) = \sum_{<\ell,d>=\alpha'} C^\ell D^\ell \phi(0).$$

This being true for all $\lambda > 0$ we must have

$$C^\ell = 0, \quad \forall \ell .$$

Let us now examine the homogeneity properties of $(pf.T_\alpha)\hat{}(x)$ when $\alpha' \in \mathbb{N}$. By using equation (3.12)

$$<(pf.T_\alpha)\hat{}(x),\phi(\lambda^{-d_1}x_1,\ldots,\lambda^{-d_n}x_n)>$$
$$= \int_0^\infty \frac{1}{p!}(\log\frac{1}{\rho} - 1\ldots - \frac{1}{p})h^{(p+1)}_{\phi(\lambda^{-d}x)}(\rho)\,d\rho \tag{3.21}$$

with $p = \alpha'$. Now

$$h^{p+1}_{\phi(\lambda^{-d}x)\hat{}(\xi)}(\rho) = \int_{\Sigma_n} T(\omega) \frac{d^{(p+1)}}{d\rho^{(p+1)}}\phi(\lambda^{-d}x)\hat{}(\rho^d\omega)\,d\omega$$

$$\phi(\lambda^{-d}x)\hat{}(\rho^d\omega) = \int_{\mathbb{R}^n} \phi(\lambda^{-d}x)e^{-<x,\rho^d\omega>}dx$$

$$= \lambda^{|d|} \int_{\mathbb{R}^n} \phi(y)e^{-<\lambda^d y,\rho^d\omega>}dy$$

with $y = \lambda^{-d}x$, $dx = \lambda^{|d|}dy$.
Therefore, $\phi(\lambda^{-d}x)\hat{}(\rho^d w) = \lambda^{|d|}\hat{\phi}((\lambda\rho)^d w)$. Putting this back to (3.21) gives

$$<(pf.T_\alpha)\hat{}(x),\phi(\lambda^{-d}x)>$$

$$= \lambda^{|d|} \int_{\Sigma_n} \int_0^\infty (\log \frac{1}{\rho} - 1 \ldots \frac{1}{p}) T(\omega) \frac{d^{(p+1)}}{d\rho^{(p+1)}} \hat{\phi}((\lambda\rho)^d \omega)\, d\rho\, d\omega .$$

By changing variable $\rho \to \rho/\lambda$ in the last integral one obtains

$$<(pf.T_\alpha)^\wedge(x), \phi(\lambda^{-d}x)>$$

$$= \lambda^{|d|+p} <(pf.T_\alpha)^\wedge(x), \phi(x)> \tag{3.22}$$

$$+ \lambda^{|d|+p} \frac{\log \lambda}{p!} \int_{\Sigma_n} T(\omega) \int_0^\infty \frac{d^{p+1}}{d\rho^{p+1}} \hat{\phi}(\rho^d \omega)\, d\rho\, d\omega .$$

Now

$$\int_0^\infty \frac{d^{p+1}}{d\rho^{(p+1)}} \hat{\phi}(\rho^d \omega)\, d\omega = -\frac{d^p}{d\rho^p} \hat{\phi}(\rho^d \omega) \Big|_{\rho=0} ,$$

$$\frac{d^p}{d\rho^p} \hat{\phi}(\rho^d \omega) = \frac{d^p}{d\rho^p} \int_{\mathbb{R}^n} \phi(x) e^{<x,\rho^d \omega>} dx \tag{3.23}$$

$$= \int_{\mathbb{R}^n} \phi(x) \frac{d^p}{d\rho^p} e^{i<x,\rho^d \omega>} dx$$

Therefore (3.23) equals

$$\int_{\mathbb{R}^n} \phi(x) \frac{d^p}{d\rho^p} (e^{i<x,\rho^d \omega>}) \Big|_{\rho=0} dx. \tag{3.24}$$

Lemma 3.2. $\frac{d^p}{d\rho^p} e^{i<x,\rho^d \omega>} \Big|_{\rho=0} = P_\omega(x)$ *is a semihomogeneous polynomial in* x *of degree* p *with respect to* $(d_1,\ldots,d_n)$.

*Proof.* As in Lemma 3.1 we use induction on the order of the differentiation. For p = 1, letting $u = <x,\rho^d \omega>$

$$\frac{d}{d\rho} e^u = \sum_{j=1}^n i(x_j \omega_j) d_j \rho^{d_j - 1} e^u . \tag{3.25}$$

The only nonvanishing terms are those whose indices are such that $d_j = 1$. Since u = 1 when p = 0 this is a semihomogeneous polynomial of degree 1. Now assume $\frac{d^\ell}{d\rho^\ell} e^u \Big|_{\rho=0}$ is a semihomogeneous polynomial for $\ell < p$. Write

$$\frac{d^p}{d\rho^p} e^u\Big|_{\rho=0} = \frac{d^{p-1}}{d\rho^{p-1}}\Big[\Big(i\sum_{j=1}^{n} x_j\omega_j d_j\rho^{d_j-1}\Big)e^u\Big]\Big|_{\rho=0}$$

$$= \sum_{s+t=p-1} \frac{1}{s!t!} \frac{d^t}{d\rho^t}\Big(i\sum_{j=1}^{n} x_j\omega_j d_j\rho^{d_j-1}\Big)\Big|_{\rho=0} \frac{d^s}{d^s} e^u\Big|_{\rho=0} \tag{3.26}$$

By induction $\frac{d^s}{d\rho^s} e^u\Big|_{\rho=0}$ is semihomogeneous of degree $s$ since $s < p$. It follows similar to the proof of Lemma 3.1 that

$$\frac{d^t}{d\rho^t}\Big(i\sum_{j=1}^{n} x_j\omega_j d_j\rho^{d_j-1}\Big)_{\rho=0} \tag{3.27}$$

is semihomogeneous of degree $t + 1$. Using Proposition 3.4 and (3.24) the second term on the right of (3.22) is

$$\lambda^{p+|d|} \int_{\mathbb{R}^n} \frac{\log\lambda}{p!} \int_{\Sigma} T(\omega)P_\omega(x)\phi(x)\,d\omega dx. \tag{3.28}$$

Now let

$$P(x) = \frac{1}{p!} \int_{\Sigma} T(\omega)P_\omega(x)\,d\omega.$$

we can write (3.28) as

$$\lambda^{|p|+d} \log\lambda \langle P(x),\phi(x)\rangle.$$

Therefore, it follows from above that we have proved the following:

Proposition 3.5. *If* $p = \alpha' \in \mathbb{N}$ *then*

$$\langle (pf.T_\alpha)^\wedge(x),\phi(\lambda^{-d}x)\rangle = \lambda^{p+|d|}\langle (pf.T_\alpha)^\wedge(x),\phi(x)\rangle$$
$$+ \lambda^{p+|d|} \log\lambda \langle P(x),\phi(x)\rangle. \tag{3.29}$$

Now consider the distribution

$$H(x) = (pf.T_\alpha)^\wedge(x) - \log\rho(x)P(x).$$

It follows from Proposition 3.5 that $H(x)$ is semihomogeneous distribution.

We know from Proposition 3.3 that $(\text{pf}.T_\alpha)\hat{}(x)$ belongs to $\Gamma^d(\mathbb{R}^n\setminus\{0\})$. Also, since $\log \rho(x) \in \Gamma^d(\mathbb{R}^n\setminus\{0\})$ and $P(x)$ is a polynomial, $H(x)$ is in $\Gamma^d(\mathbb{R}^n\setminus\{0\})$. Thus $H(x)$ is the sum of a function in $\Gamma^d(\mathbb{R}^n\setminus\{0\})$ and a distribution supported at the origin. The argument used in the paragraph following the proof of Proposition 3.3 shows that this distribution must be zero.

## 3.4 REGULARITY OF THE KERNELS

We now apply the results of Section 3.3 to show that $K_\nu(x,t)$ defined by 3.4 belong to $\Gamma^d(\mathbb{R}^n\setminus\{0\})$, as well as give their representation. Now

$$H_\nu(\xi,t) = \begin{vmatrix} Q_1(\xi,\tau_1(\xi)) & \cdots & Q_1(\xi,\tau_\mu(\xi)) \\ \vdots & & \\ e^{it\tau_1(\xi)} & \cdots & e^{it\tau_\mu(\xi)} \\ & & \\ Q_\mu(\xi,\tau_1(\xi)) & \cdots & Q_\mu(\xi,\tau_\mu(\xi)) \end{vmatrix} \cdot (C(\xi))^{-1} \tag{3.30}$$

which we can write as

$$H_\nu(\xi,t) = \sum_{j=1}^{n} f_{j,\nu}(\xi)e^{it\tau_j(\xi)}$$

where

$$f_{j,\nu}(\xi) = \frac{(-1)^{\nu+j}\det M_{\nu,j}(\xi)}{C(\xi)}$$

and $M_{\nu,j}(\xi)$ is the $(\nu,j)$ minor in the determinant of (3.30). Since $\tau_j(\lambda^{d_1}\xi_1,\ldots,\lambda^{d_n}\xi_n) = \lambda^{d_{n+1}}\tau_j(\xi)$ and $C(\xi)$ is semihomogeneous of degree

$$n_1 + n_2 + \ldots + n_\mu - d_{n+1}(1+2+\ldots+(\mu-1))$$

$f_{j,\nu}(\xi)$ is semihomogeneous of degree $k_\nu$:

$$\begin{aligned} k_\nu &= (n_1+n_2+\ldots+\hat{n}_\nu+\ldots+n_\mu) \\ &\quad -(n_1+\ldots+n_\mu-d_{n+1}(1+2+\ldots+(\mu-1))) \\ &= -n_\nu + d_{n+1}\left(\frac{\mu(\mu-1)}{2}\right). \end{aligned} \tag{3.31}$$

We thus can apply the methods of the previous section to define $(\text{pf}.T_{j,\nu})\hat{}(x)$. Since it makes sense to multiply a distribution by a $C^\infty$ function we define

$$K_{j,\nu}(x,t) = (e^{it\tau_j(\xi)}\,\text{pf}.f_{j,\nu}(\xi))\hat{}(x)$$

and thus

$$K_\nu(x,t) = F^{-1}H_\nu(\xi,t) = \sum_{j=1}^{\mu} K_{j,\nu}(x,t)$$

Now if $k'_\nu \notin \mathbb{N}$ by (3.9) for $\phi(x,t) \in C_c^\infty(\mathbb{R}_+^{n+1})$

$$\begin{aligned}&\langle K_{j,\nu}(x,t),\phi(\lambda^{-d_1}x_1,\ldots,\lambda^{-d_n}x_n,\lambda^{-d_{n+1}}t)\rangle \\ &= \lambda^{|d|}[k'_\nu(k'_\nu-1)\ldots(k'_\nu-p)]^{-1}\int_0^\infty\int_{\Sigma_n}\int_-^\infty f_{j,\nu}(\omega)\rho^{-k'_\nu+\rho} \times \\ &\qquad \frac{d^{p+1}}{d\rho^{p+1}}\, e^{it\tau_j(\rho^d\omega)}\hat{\phi}((\lambda\rho)^d\omega,\lambda^{-d_{n+1}}t)]\,d\rho\, d\omega\, dt.\end{aligned} \tag{3.32}$$

Making the change of variable

$$r = \lambda\rho, \quad s = \lambda^{-d_{n+1}}t,$$

(3.32) equals

$$\begin{aligned}&\lambda^{d_{n+1}}\lambda^{-1}\lambda^{p+1}\lambda^{|d|}[k'_\nu\ldots(k'_\nu-p)]^{-1}\int_0^\infty\int_{\Sigma_n}\int_0^\infty f_{j,\nu}(\omega)\left(\frac{r}{\lambda}\right)^{-k'_\nu+p} \times \\ &\qquad \frac{d^{p+1}}{dr^{p+1}}\,[e^{i\lambda^{d_{n+1}}s\tau_j((\frac{r}{\lambda})^d\omega)}\hat{\phi}(r^d,s)\,d\rho\, d\omega\, ds\end{aligned}$$

Since $\tau_j((\frac{r}{\lambda})^d\omega) = \lambda^{-d_{n+1}}\tau_j(r^d\omega)$ the last integral equals

$$\lambda^{|d|+d_{n+1}+k'_\nu}\langle K_{j,\nu}(x,t),\phi(x,t)\rangle \tag{3.33}$$

This shows that $K_{j,\nu}(x,t)$ is a semihomogeneous distribution of degree $k'_\nu$ with respect to $(d_1,\ldots,d_{n+1})$.

If $k'_\nu \in \mathbb{N}$ then one can show, as in the previous section, that

$$K_j(x,t) = H_j(x,t) + \log \rho(x,t)P(x,t)$$

where $H_j(x,t)$ is semihomogeneous of degree $k'_\nu$ and $P(x,t)$ is a semihomogeneous polynomial both with respect to $(d_1,\ldots,d_{n+1})$.

Furthermore, these functions belong to $\Gamma^d(\mathbb{R}^{n+1}\setminus\{0\})$ in the x variable since we took the Fourier transform in this variable. From equation (2.74) we see that if K is a compact subset of $\mathbb{R}_+^{n+1}\setminus\{0\}$.

$$\sup_{(x,t)\in K} |D_t^\beta H_\nu(\cdot,t)(x)| \le C^{|\beta|+1}$$

for $t \ge 0$, $0 \le |\beta| \le \sigma-1$ where $\sigma$ is the transverse order of $P(D,D_t)$. Using a similar induction argument as at the end of Section 2.10 we conclude that $K_\nu(x,t)$ is Gevrey in the t variable and hence

$$K_\nu(x,t) \in \Gamma^{(d_1,\ldots,d_{n+1})}(\mathbb{R}_+^{n+1}\setminus\{0\}).$$

One can perform the same analysis on $K_0(x,t)$ and thus we have proved the following.

Theorem 3.2. *If $(d_1,\ldots,d_{n+1})$ is an (n+1)-tuple of positive integers then $(P;Q_1,\ldots,Q_\mu)$ defines a semielliptic boundary-value problem if and only if there exists a fundamental solution $K_0(x,t),\ldots,K_\mu(x,t)$ where $K_j(x,t)$ are either semihomogeneous or of the form*

$$K_j(x,t) = H_j(x,t) + \rho(x,t)P_j(x,t)$$

*with $H_j(x,t)$ semihomogeneous distribution and $P_j(x,t)$ a semihomogeneous polynomial. In either case $K_j(x,t) \in \Gamma^d(\mathbb{R}_+^{n+1}\setminus\{0\})$ for all j.*

## 3.5 SEMIELLIPTIC PROBLEMS

We now consider the more general type of semielliptic boundary-value problem discussed in Section 3.1. We wish to prove Theorem (3.1).

Theorem 3.1

*(1) If $C^0(\xi) \ne 0$ for all $\xi \in \mathbb{R}^n\setminus\{0\}$ then $(P;Q_1,\ldots,Q_\mu)$ defines a regular semielliptic boundary-value problem in $\Omega\cup\omega$.*

*(2) Conversely, if* $(P;Q_1,\dots,Q_\mu)$ *defines a regular semielliptic boundary-value problem in* $\Omega \cup \omega$ *then either* $C^0(\xi) \equiv 0$ *or* $C^0(\xi) \neq 0$ *for* $\xi \in \mathbb{R}^n \setminus \{0\}$.

Before proceeding with the proof of Theorem 3.1 let us make several remarks about the roots of $P(\xi,\tau) = 0$ and $P^0(\xi,\tau) = 0$. If $\tau^0(\xi)$ denotes a root of the second equation, it is the limit as $t \to +\infty$ of

$$t^{-d_{n+1}}\tau(t^{d_1}\xi_1,\dots,t^{d_n}\xi_n), \tag{3.38}$$

where $\tau(\xi)$ is a root of the first equation. Also, $\tau^0(\xi)$ is semihomogeneous of degree $d_{n+1}$ with respect to $(d_1,\dots,d_n)$.

Also, observe that

$$\lim_{t\to+\infty} t^{-N} C(t^{d_1}\zeta_1,\dots,t^{d_n}\zeta_n) = C^0(\zeta), \tag{3.39}$$

where $C(\zeta)$ is the characteristic function of $(P;Q_1,\dots,Q_\mu)$. This follows easily from the definition of the characteristic functions, the semihomogeneity of $P$ and $Q_\nu$, $1 \le \nu \le \mu$, and the previous remark (3.38).

Next suppose that $C^0(\xi) \neq 0$, for all $\xi \in \mathbb{R}^n$, $\xi \neq 0$. Then for some small $\varepsilon > 0$ and some constant $C_1 > 0$ we have

$$\begin{gathered} |C^0(\zeta)| \ge C_1, \quad \text{if} \quad \zeta \in C^n, \\ [\zeta]_{d'} = 1 \quad \text{and} \\ [\operatorname{Im} \zeta]_{d'} \le \varepsilon[\operatorname{Re} \zeta]_{d'}. \end{gathered} \tag{3.40}$$

Indeed, the assumption $C^0(\xi) \neq 0$, $\forall\, \xi \in \mathbb{R}^n$, $\xi \neq 0$, implies the existence of a constant $C_1 > 0$ such that

$$C^0(\xi) \ge C_1, \quad \forall\, \xi \in \mathbb{R}^n \text{ such that } [\xi]_{d'} = 1.$$

But the last inequality can be extended to a small complex neighborhood of the set

$$\{\xi \in \mathbb{R}^n : [\xi]_{d'} = 1\}$$

thus yielding (3.40).

On the other hand, (3.40) is equivalent to

$$|C^0(\zeta)| \ge C_1[\zeta]^N_{d'}, \tag{3.41}$$

if $\zeta \in C^n$ and $[\mathrm{Im}\ \zeta]_{d'} \leq \varepsilon[\mathrm{Re}\ \zeta]_{d'}$.

*Proof of Theorem 3.1.* (a) (1 => 2). From our assumption (3.41) holds. We claim that

$$[\zeta]_{d'} \geq C \quad \text{and} \quad [\mathrm{Im}\ \zeta]_{d'} \leq \varepsilon[\mathrm{Re}\ \zeta]_{d'} \tag{3.42}$$

with C a large constant, imply $C(\zeta) \neq 0$. To see this write $C(\zeta) = C^0(\zeta) + C(\zeta)-C^0(\zeta)$. We then have

$$|C(\zeta)| \geq |C^0(\zeta)|-|C(\zeta)-C^0(\zeta)| \geq C_1\ [\zeta]^N_{d'}-|C(\zeta)-C^0(\zeta)|. \tag{3.43}$$

Let

$$t = [\zeta]_{d'} \text{ and } \eta_j = \zeta_j/t^{d_j}.$$

From (3.39), given $\delta > 0$, there is a constant $C > 0$ such that

$$|t^{-N}C(\zeta)-C^0(\eta)| \leq \delta$$

provided that $t \geq c$. Hence

$$|C(\zeta)-C^0(\zeta)| \leq \delta t^N, \quad \forall\ t \geq c.$$

Replacing this in (3.43) we get

$$|C(\zeta)| \geq (C_1-\delta)t^N, \quad \forall\ t \geq c.$$

This implies (3.42) provided $\delta$ is sufficiently small. Finally, let $\bar{d} = \max\limits_{1\leq j\leq n} (d_j)$ and choose $\varepsilon_1 < 1$ so that $\frac{1}{n}\left(\frac{1}{\varepsilon_1}\right)^{1/\bar{d}} > \frac{1}{\varepsilon}$. Take $M \geq 1$ so large that $M \geq \max\left(\frac{1}{\varepsilon_1},c\right)$ and that the set

$$\{\zeta \in C^n : |R(\zeta)|_{d'} \geq M(1+|\mathrm{Im}\ \zeta|)\} \tag{3.44}$$

is contained in $A$. If $\zeta \in C^n$ belongs to the set (3.44) then it is obvious that

$$[\zeta]_{d'} \geq C.$$

On the other hand we also have

$$[\mathrm{Re}\ \zeta]_{d'} \geq \frac{1}{\varepsilon_1}|\mathrm{Im}\ \zeta_\ell|, \quad 1 \leq \ell \leq n.$$

Raising both sides of this inequality to the $d_\ell^{-1}$ power, taking into account that $d_\ell \geq 1$, $M \geq 1$ and the choice of $\varepsilon_1$, we get:

$$[\mathrm{Re}\ \zeta]_{d'} \geq \frac{1}{n}\left(\frac{1}{\varepsilon_1}\right)^{1/\bar{d}}[\mathrm{Im}\ \zeta]_{d'} \geq \frac{1}{\varepsilon}\ [\mathrm{Im}\ \zeta]_{d'}.$$

Thus if $\zeta$ belongs to the set (3.44) it satisfies (3.42) which implies that $C(\zeta) \neq 0$ but then it follows from Chapter 2 that $(P;Q_1,\ldots,Q_\mu)$ is a regular semielliptic boundary-value problem since this was precisely the condition needed to construct the parametrix of $(P;Q_1,\ldots,Q_\mu)$.

(b) (2 => 1). Let $\xi \in \mathbb{R}^n$, $\xi \neq 0$, be such that $C^0(\xi) = 0$. Let be a real vector and $\omega$ be any complex number. Define

$$\sigma_j = \begin{cases} 0 & \text{if } \xi_j = 0 \\ w & \text{if } \xi_j \neq 0 \text{ and } d_j = 1 \\ |\mathrm{Re}\ w|^{d_j} & \text{if } \xi_j \neq 0 \text{ and } d_j > 1 \end{cases} \tag{3.45}$$

and

$$\zeta_j = s^{d_j}(\xi_j+\sigma_j\eta_j), \quad 1 \leq j \leq n. \tag{3.46}$$

From our semihomogeneity assumption it follows that

$$\lim_{s\to+\infty} s^{-M}\, C(\zeta_1,\ldots,\zeta_n) = C^0(\xi_1+\sigma_1\eta_1,\ldots,\xi_n+\sigma_n\eta_n).$$

We claim that for suitably chosen w, $C^0(\xi+\sigma\eta)$ is either never zero or identically zero. First let us find the appropriate condition on w.

Case (a). The index j is such that $\xi_j \neq 0$ and $d_j = 1$. In this case we have

$$\mathrm{Re}\ \zeta_j = s(\xi_j+\mathrm{Re}\ w\ \eta_j), \quad \mathrm{Im}\ \zeta_j = s\ \mathrm{Im}\ w\ \eta_j.$$

Then

$$|\mathrm{Re}\ \zeta_j| = s|\zeta_j+\mathrm{Re}\ w\ \eta_j| \geq s(|\xi_j|-|\mathrm{Re}\ w|\ |\eta_j|)$$

and

$$|\mathrm{Im}\ \zeta_j| = s|\mathrm{Im}\ w||\eta_j|.$$

For $|w|$ small and s large, we impose

$$s(|\xi_j|-|\text{Re } w||\eta_j|) \geq M(1+s|\text{Im } w||\eta_j|). \tag{3.47}$$

Case (b). The index j is such that $\xi_j \neq 0$ and $d_j > 1$. In this case we have

$$\text{Re } \zeta_j = s^{d_j}(\xi_j+|\text{Re } w|^{d_j}\eta_j) \text{ and } \text{Im } \zeta_j = 0.$$

It follows, since $d_j > 1$, that

$$\begin{aligned}|\text{Re } \zeta_j|^{1/d_j} &= s|\xi_j+|\text{Re } w|^{d_j}\eta_j|^{1/d_j} \\ &\geq s(|\xi_j|^{1/d_j}-|\text{Re } w||\eta_j|)^{1/d_j}).\end{aligned}$$

We now impose for small $|w|$ and large s

$$s|\xi_j|^{1/d_j}-|\text{Re } w||\eta_j|^{1/d_j} \geq M(1+s|\text{Im } w||\eta_j|). \tag{3.48}$$

The inequalities (3.47) and (3.48) imply

$$\begin{aligned}[\text{Re}]_{d'} &\geq s([\xi]_{d'}-|\text{Re } w|[\eta_j]_{d'}) \\ &\geq M(1+s|\text{Im } w|\sum_{j=1}^{n}|\eta_j|) \\ &= M(1+\sum_{j=1}^{n}|\text{Im } \zeta_j|) \geq M(1+|\text{Im } \zeta|).\end{aligned}$$

That is $\zeta \in A$, $C(\zeta)$ is analytic and $C(\zeta) \neq 0$. On the other hand, (3.47) is implied by

$$|w| \leq \frac{|\xi_j|-Ms^{-1}}{(1+M)|\eta_j|} \tag{3.49}$$

and (3.48) is implied by

$$|w| \leq \frac{|\xi_j|^{1/d_j}-Ms^{-1}}{(1+M)|\eta_j|^{1/d_j}}, \text{ provided } |\eta_j| < 1. \tag{3.50}$$

We now make the choice of w such that

$$|w| \leq \min_{j,k} \left(\frac{|\xi_j|}{(1+M)|\eta_j|}, \frac{|\xi_k|^{1/d_k}}{(1+M)|\eta_k|^{1/d_k}}\right) = \delta. \tag{3.51}$$

We claim that for this choice of w, the $C^0(\xi+\sigma\eta)$ is either identically zero or never zero. Indeed, suppose that for some $|w_0| < \delta$ we had $C^0(\xi+\sigma^0\eta) = 0$. Then it would be possible to find s sufficiently large so that (3.49) and (3.50) hold for $w_0$ and that

$$s^{-M}|C(\zeta_1^0,\ldots,\zeta_n^0)| < \varepsilon$$

for arbitrary $\varepsilon > 0$. By continuity, this would imply

$$C^0(\xi+\sigma\eta) = 0 \quad \text{with } \sigma = (\sigma_1,\ldots,\sigma_n)$$

associated to w in a suitable neighborhood of $w_0$. But by assumption, $C^0(\xi+\sigma\eta) = 0$ when $w = 0$, and thus $C^0(\xi+\sigma\eta) = 0$ in the circle $|w| < \delta$. Now if we take

$$\begin{cases} |\eta_j| < \dfrac{|\xi_j|}{1+M}, & \forall_j \text{ such that } \xi_j \neq 0 \text{ and } d_j = 1. \\ |\eta_j| < \min\left(1, \dfrac{||\xi_j|}{(1+M)^{d_j}}\right), & \forall_j \text{ such that } \xi_j \neq 0 \text{ and } d_j > 1. \\ |\eta_j| \text{ arbitrary for all other } j. \end{cases} \tag{3.52}$$

the circle (3.51) contains $w = 1$. It then follows that $C^0(\xi+\eta) = 0$ for $\eta$ in the neighborhood of zero defined by (3.52). Since $C^0(\zeta)$ is analytic there it must vanish identically. Q.E.D.

BIBLIOGRAPHY

1. R. Artino, Gevrey classes and hypoelliptic boundary-value problems, Pac. J. of Math., Vol. 63, No. 1, (1976), 1-21.

2. R. Artino, On semielliptic boundary-value problems, J. Math. Anal. Appl. 42, 3 (1973), 610-626.

3. R. Artino and J. Barros-Neto, Regular semielliptic boundary-value problems, J. Math. Anal. Appl. 61, 1 (1977), 40-57.

4. J. Barros-Neto, Kernels associated to general elliptic problems, J. of Funct. Anal., No. 2 (1969), 172-192.

5. J. Barros-Neto, On Poisson's kernels, Anais da Academia Brasileira de Ciencias, 42, No. 1 (1970), 1-3.

6. J. Barros-Neto, The parametrix of a regular hypoelliptic boundary-value problem, Ann. Sc. Norm. Sup. Pisa 26, Fasc. 1 (1972), 247-268.

7. J. Barros-Neto, On regular hypoelliptic boundary-value problems, J. Math. Anal. Appl. 41, No. 2 (1973), 508-530.

8. J. Barros-Neto, An Introduction to the Theory of Distributions, Marcel Dekker, New York, 1973.

9. L. Ehrenpreis, Solutions of some problems of division, I. Am. J. Math. 76 (1954), 883-903.

10. E.B. Fabes and N.M. Riviere, Singular integrals with mixed homogeneity, Studia Math. 27 (1966), 19-38.

11. L. Hormander, On the regularity of the solutions of boundary problems, Acta Math. 99 (1958), 225-264.

12. L. Hormander, Linear Partial Differential Operators, Springer-Verlag, Berlin, 1963.

13. B. Malgrange, Existence et approximation des solutions des equations aux derivees partielles et des equations de convolution, Ann. Inst. Fourier, Grenoble 6 (1955-56), 271-355.

14. L. Schwartz, Theorie des Distributions, I, II, Hermann and Cie, Paris, 1957.

15. L. Schwartz, Seminaire 1954/55, Equations aux Derivees Partielles, Paris, 1955.

16. F. Treves, Linear Partial Differential Equations with Constant Coefficients, Gordon and Breach, New York, 1966.

# INDEX